A THIN COSMIC RAIN

A THIN COSMIC RAIN

PARTICLES FROM OUTER SPACE

MICHAEL W. FRIEDLANDER

HARVARD UNIVERSITY PRESS
CAMBRIDGE, MASSACHUSETTS, AND LONDON, ENGLAND 2000

An earlier version of this book was published in 1989 as *Cosmic Rays*.

Library of Congress Cataloging-in-Publication Data
Friedlander, Michael W.
A thin cosmic rain : particles from outer space / Michael W. Friedlander.
    p. cm.
Rev. ed. of: Cosmic rays, 1989.
Includes bibliographical references and index.
ISBN 0-674-00288-1 (clothbound : alk. paper)
1. Cosmic rays. I. Friedlander, Michael W. Cosmic rays. II. Title.
QC485 .F75 2000
539.7'223—dc21   00-039594

For Jessica

CONTENTS

A THIN COSMIC RAIN

1

"COMING OUT OF SPACE and incident on the high atmosphere, there is a thin rain of charged particles known as the primary cosmic radiation." With these words, Cecil Powell began his lecture at the ceremonies in Stockholm when accepting the Nobel Prize for Physics in 1950. Powell was being honored for the improvements he had made in a method of detecting high-speed atomic particles and for his discovery among the cosmic rays of mesons, minute but radioactive constituents of matter.

What are cosmic rays? What kind of radiation are they, and why cosmic? Where do cosmic rays come from, and what do they do? Are they harmful to us? Now, almost a hundred years after their discovery, we know that the cosmic rays Powell described consist mostly of high-speed protons, the nuclei of hydrogen atoms, with about 9 percent helium and heavier nuclei and a small percentage of electrons. The Earth is continually bombarded by these particles, which collide with atoms in the stratosphere to produce more particles that can be detected throughout the atmosphere and sometimes deep underground.

We have identified possible sources of this cosmic rain. Supernovas, the explosive late stage in the evolution of stars much more massive than our sun, are known to provide many fast particles, and they seem to meet many of the requirements for sources of cosmic rays. Most of the cosmic rays probably originate within our own (Milky Way) galaxy, but a very few particles—and only the most energetic—appear to be coming from very much farther away. The general outlines of this picture are coming into a fuzzy focus, but many of the important details remain to be clarified. We know that some cosmic rays come from the sun. We also know that the sun and the Earth's magnetic field combine to exert a major influence on the slowest cosmic rays. Cosmic rays pass through us at all times and give us small but unavoidable doses of radiation.

Any mention of radiation raises public concerns today, after Hiroshima and Nagasaki, and the accidents at the Three Mile Island and Chernobyl nuclear power reactors. This was not always the case, however; in the early years of this century, radiation was seen not as hazardous but rather as something mysterious, even beneficial. Scientific discoveries and research had identified several different but sometimes related forms of radiation, and medical science was beginning to see some almost-magical prospects for the use of X-rays in diagnosis and therapy. There was an emotional public response when Marie Curie toured the United States in 1921 to raise funds for the purchase of 1 gram of radium for her research institute. Arthur Compton won the 1927 Nobel Prize for his research with X-rays, which provided important support for the new quantum theory. Radiation was a hot subject, and people were fascinated by cosmic rays, even though (or perhaps because) they were enigmatic, from somewhere out in the cosmos.

In many ways, cosmic rays still retain this aura of mystery. As we will see, cosmic rays have been and remain central to a remarkably diverse range of research studies, from the smallest imaginable scale involving the composition of the atomic nucleus to the vast scale of galactic dimensions. Cosmic rays have been used to probe the magnetic field surrounding the Earth and the regions between the stars. The study of cosmic rays has stimulated the development of techniques for the detection of radiation as well as research into ways in which radiation affects atoms and molecules, in matter both living and inert. Cosmic rays are studied from satellites and space probes, from balloons in the stratosphere, with high-altitude airplanes, and with giant detectors 2 miles beneath the Earth's surface. The fundamental nature of many discoveries in

cosmic ray research has been recognized through the award of a dispropor-
tionately large number of Nobel Prizes.

How did all this activity start? The first intimations of the presence of cosmic rays came quite unexpectedly, around the turn of the century, during the
golden days of research into radioactivity. Sensitive instruments recorded radiation even when there was no radium or uranium nearby, and it was soon
found that this new radiation did not originate on the Earth but came from
somewhere else. The pursuit of this cosmic radiation has led us first to high
mountains, then to the stratosphere, and finally to great distances using space
probes so that locally produced effects could be eliminated and our attention
focused on the primary rays themselves.

For about 20 years cosmic ray research was the only arena for the discovery
of new subatomic particles. As we now know, these particles emerge from
collisions between nuclei traveling at more than half the speed of light. Some
of the collision energy reappears as the mass of the new particles, following
Einstein's famous formula $E = mc^2$. The first of the new particles turned up
among the cosmic rays in the early 1930s, and their numbers multiplied after
World War II, when cosmic ray research expanded rapidly. The way to the
modern era of elementary particle physics was paved during those years, and
cosmic rays were effectively displaced as the source of new particles only with
the advent of the giant accelerators in the mid-1950s. Yet even the most energetic particles from these accelerators are still many millions of times less energetic than some that can be found (but rarely and uncontrollably) among
the cosmic rays.

At the same time as our knowledge of the primary cosmic rays was growing
in the years after 1950, radio astronomy was becoming firmly established. In
addition, with observations from above the atmosphere, the scope of astronomy broadened further to include the infrared, ultraviolet, and X-ray regions
of the spectrum. It has been found that much of this radiation emerges from
processes that also produce fast particles, and cosmic rays are now an important component of contemporary astrophysics, with a professional journal,
Astroparticle Physics, devoted to them.

Although the main focus of research on cosmic rays is now their nature and origin, we have also discovered that these high-speed particles leave clear signatures on Earth as they penetrate into the atmosphere and collide with atoms. Nuclear debris from cosmic ray collisions can be traced from the atmosphere along some unexpected but fascinating paths. For example, radiocarbon dating, so important in archaeology, owes its existence to cosmic rays. I will explore some of these topics later in the book, but first I will retrace the historical path through the various discoveries, some the results of design but others quite accidental, that led to the detection of cosmic rays.

The late nineteenth century was a time of consolidation in physics. Gravity was thought to be well understood, and James Clerk Maxwell's new theory of electromagnetism tied together many previously separated topics. Although some puzzles had not been solved, there was a general feeling that the major discoveries had been made and all that remained was to pursue various quantities to their next decimal places.

Among the topics of research was the electrical conductivity of gases, which was thought to be interesting, but scarcely revolutionary. As so often happens in science, however, this study began to follow unpredictable directions. The incandescent electric lamp provided the first of these surprises. This device contains a small, tightly coiled metal filament that emits light when an electric current flows through it. The filament also boils off electrons, but these negatively charged particles soon return to the filament, which would otherwise build up a positive charge. If, though, the glass container enclosing the filament has a metal probe inside, and if a high-voltage battery is connected at its negative terminal to the filament and at its positive terminal to the probe, electrons from the filament will travel directly through the gas to the probe. In this arrangement, the filament, or cathode, and the other probe (or electrode) the anode, are the major components of the cathode-ray tube (or CRT), used until recently in every television set.

The CRT, developed in the 1880s, became a major tool for studying electrical conduction through gases because the stream of cathode rays could be deflected by electric and magnetic fields. In the late nineteenth century, the nature of these rays was revealed by J. J. Thomson, of the Cavendish Laboratory at Cambridge University. Until that time, as Thomson put it, "the most diverse opinions [were] held as to these rays," but his 1897 paper described quantitatively how cathode rays—now known as electrons—behave like particles. Thomson's discovery was a turning point in modern physics.

As electrons travel through the gas in the CRT, as shown in Figure 1.1, they knock other electrons off some of the gas atoms, which then become positively charged ions. In their turn, these ions are accelerated by the high voltage and constitute the positive rays also studied by Thomson. From this research came the discovery of isotopes, atoms of the same element with the same chemical properties (in the way in which they combine with other atoms to form molecules) but different masses.

From the discovery of radioactivity through the invention of the cyclotron in the 1930s to the modern era of nuclear research, a central thread has been the development of ways to detect and identify high-speed particles. Because most cosmic rays are electrons and nuclei and travel with speeds close to the speed of light, their detection and speed measurement often draw on methods devised for nuclear physics experiments at giant accelerators.

The turn of the century was a particularly exciting time for uncovering

Figure 1.1. J. J. Thomson's cathode-ray tube. The negative terminal of the high voltage supply is connected to the cathode (C), and the positive terminal to the anode (A). Accelerated electrons pass through slits in the anode and another screen (B) and can travel until they strike the end of the tube (S), which is coated with a fluorescent material. The cathode-ray beam can be deflected by application of a suitable voltage across the additional plates (D and E) or by a magnet.

the intricacies of atomic structure, the properties of electrons, and the different types of ions. Even though Maxwell's new theory of electromagnetism seemed to provide many of the answers, the nature of light and its emission continued to be investigated. In the 1890s, Wilhelm Conrad Roentgen at the University of Würzburg was studying the phenomena of phosphorescence and fluorescence, in which materials emit light on their own. When using a cathode-ray tube, Roentgen noted that the glass fluoresced where the cathode rays hit. Fluorescence was also observed near the cathode-ray tube when a glass plate (or screen) with a sensitized coating was used. One day in November 1895, Roentgen found that his fluorescent plate glowed even when at a considerable distance from the cathode-ray tube and even when the tube was completely covered by black cardboard. Clearly, some radiation was reaching his plate and penetrating the paper and other materials that covered the plate. Putting his hand between the plate and the tube, he could see the outline of his bones on the fluorescent screen. Roentgen had discovered X-rays. The medical use was recognized at once. Roentgen reported his discovery at the weekly meeting of the Academie des Sciences in Paris on 20 January 1896, and in 1902 he received the first Nobel Prize for Physics.

Worldwide interest was aroused. Henri Becquerel, who heard Roentgen's first presentation at the Academie des Sciences, was prompted to resume his own work on fluorescence. Less than a month later, while using compounds of uranium, Becquerel noticed that photographic materials became fogged, even when they were fully wrapped to protect them against fluorescence from the chemicals he was testing or against sunlight. It soon became clear that the uranium itself was emitting some sort of radiation without the stimulation of sunlight. Over the next few years this property of radioactivity was found in thorium and then, by Pierre and Marie Sklodowska Curie, in previously unidentified chemical elements that they named radium and polonium.

X-rays and the radioactive emanations ionize gases, enabling the gases to conduct electricity. This effect provides the means for detecting and measuring radiation, and electroscopes and electrometers came into wide use in those early days when X-rays and radioactive materials were being explored with such excitement. Two simple types of electroscope, shown in Figure 1.2, measure the intensity of radiation through its effect on a stored electric charge. As radioactive elements came to be identified, it was found that each element displayed a radioactive power that decreased over time. Each substance took a different amount of time for its activity to decrease to half of the starting value. This half-life, an inherent property of each radioactive element or isotope, is unaffected by pressure, temperature, or chemical combination.

As researchers puzzled over the newly discovered radiations, they encountered a troubling experimental fact. No matter how well they made their electroscopes and no matter how good the insulation, the electric charge continued to leak away, even when there was no obvious nearby source of X-rays or radioactivity. More strenuous efforts reduced the rate of leakage but could not completely eliminate it—not even when the electroscope was surrounded by more than 2 inches of lead. Whatever was responsible for this leakage had a remarkably penetrating power.

There followed numerous attempts to measure the rate of residual leakage in different locations. To reduce the effects of local sources of radiation in the ground, for example, electroscopes were carried to the tops of tall buildings. The inconclusive results were attributed to radioactivity from building materi-

Figure 1.2. Electroscopes. The type on the left, with two gold leaves, is still often used for lecture demonstrations. In the Wulf electroscope (on the right), two wires are used instead of the gold leaves. When an electroscope is given an electric charge, the leaves or wires repel each other and stand apart. Radiation can ionize some of the air in the electroscope and allow the charge to leak away, as is shown by the wires or leaves slowly coming back together.

als. In 1910 Father Thomas Wulf took his electroscope about 900 feet up the Eiffel Tower and observed a reduction of 64 percent in the leakage rate. Wulf had expected that radiation from the ground would have decreased by far more than this because of absorption in the air, and he deduced that radiation from the ground (gradually decreasing with altitude) was probably competing with radiation coming down through the atmosphere.

The obvious next step was to go to greater heights, and several investigators did just that, using balloons. An Austrian physicist, Victor Hess, starting in 1911, was the first to produce decisive results from balloon flights in which he ascended along with his electroscopes (Figure 1.3). Hess noted that the intensity of the ionizing radiation first decreased as the balloons went up, but by about 5000 feet the radiation was definitely more intense than at sea level. By the maximum height of 17,500 feet, the radiation had increased several times. Hess asserted that the explanation was "an extra-terrestrial source of penetrating radiation."

Further flights, notably by Werner Kolhörster, of the Physikalisch-Technische Reichsanstalt in Berlin, and later by the Americans Robert Millikan and Ira S. Bowen, confirmed the findings. By now the balloon altitudes were great enough that the electrometers needed to be operated by remote control and the measurements recorded automatically. After his first measurements, however, Millikan concluded that "there exists no radiation of cosmic origin" because even during balloon flights over San Antonio, Texas, the radiation seemed to be "not more than 25% of that found by European observers" (p. 360). We now know that Millikan's findings of low intensities, though correct, were produced by the Earth's magnetic field, which is very different over Texas than it is over Europe. Millikan, a man who made major contributions to physics, was given to strongly held views, sometimes erroneous. In this case, he reversed his judgment after making further high-altitude measurements. By 1926 Millikan finally came to agree that the radiation did not originate on the Earth, and he introduced the name "cosmic rays."

From these early observations to the sophisticated and wide-ranging research of today stretches a tangled trail of puzzles, solutions, and new puzzles, pursued with great technical ingenuity and with frequent shifts in emphasis. In

Figure 1.3. Victor Hess, discoverer of the cosmic rays, after his 1912 balloon flight that reached an altitude of 17,500 feet. (Photograph courtesy of Martin A. Pomerantz.)

later chapters I review the current state of each of the main areas of cosmic ray research; here, I present a brief overview that relates the different areas.

It has now been firmly established that the cosmic rays fall into two broad categories. One class, in the minority, comes from the sun. Increases in cosmic ray intensity are detected in conjunction with large solar flares—giant eruptions that can be seen (by telescope) on the face of the sun. But most cosmic rays come from much farther away, as is shown by the relatively minor changes in cosmic ray flux between day and night. There is overwhelming evidence that almost all of these cosmic rays originate in our own galaxy, a region that contains about 100,000,000,000 stars, including our sun and solar system. Our galaxy is about 100,000 light years in diameter, and we are located approximately on its equatorial plane, about halfway from the center. Stars and dust are mostly concentrated in the equatorial region of the galaxy. We call this region the Milky Way, and we see it sweeping as a glowing band across the sky when we look up at night from places that are free of air pollution and bright street lights.

Within the galaxy, the sources of cosmic rays have still not been finally identified. Many, perhaps most, come from unusual stars such as supernovas, which have exploded and hurled their debris out into interstellar space. Just how and where the rapid acceleration of the particles takes place is not settled. Yet many measurements of radiation from supernova remnants clearly display the presence of high-speed electrons. The best known of these supernova remnants (SNR) is the Crab nebula in the constellation Taurus. So well has this object been studied that, as the theorist Geoffrey Burbidge has put it, astrophysics at one time consisted of the Crab nebula and everything else. The explosion of the Crab was well documented in 1054 A.D. by astronomers in China, Korea, and Japan. It was a spectacular event. The Crab was brighter than any star in the night sky—comparable to Venus, as the Chinese noted—so bright that it could be seen in daylight for 23 days. By now, the Crab has been observed in all parts of the electromagnetic spectrum, from radio to gamma rays.

It is interesting to note that, in spite of its extreme brightness, no record of the Crab's explosion has been found in Europe. One explanation is that Christian belief held that the heavens were eternally perfect, unchanging. Comets and meteors were clearly transient but were considered to be atmospheric happenings, and other changes in the heavens were simply ignored as though they had not happened. The first recorded supernova report in the West came in 1572, another in 1604, by which time the celestial beliefs of the Church were being challenged.

Understanding events like the Crab outburst requires a knowledge of the nuclear physics of the fusion processes that take place deep within stars, providing the energy that keep the stars glowing. This knowledge, in turn, depends on laboratory studies of nuclei and their methods of interaction. Nuclear astrophysics has become a major research area, and two of its pioneers, Subramanyan Chandrasekhar of the University of Chicago and Willy Fowler of the California Institute of Technology (Cal Tech), shared the Nobel Prize for Physics in 1984.

The vast majority of the nonsolar cosmic rays originate in our galaxy, but growing evidence indicates that some of the very highest energy particles come to us from outside—presumably originating in the myriad distant galaxies that telescopes have revealed. This evidence, coming from several research groups, raises yet more questions, and I will return to it in Chapter 7.

As cosmic rays approach the Earth, they encounter the Earth's magnetic field. The Earth's behavior as a giant magnet has been known for centuries; it provides the basis for the simple magnetic compass. The paths of electrically charged particles are deflected in a magnetic field, and near the Earth they become highly convoluted, depending in detail on each particle's speed and arrival direction. As a result, the numbers and speeds of the cosmic rays close to Earth are found to vary with latitude and will not be the same as those measured far away. In order to understand the properties of the primary cosmic rays, we must make corrections to our local measurements. Between the 1930s and the 1960s, many cosmic ray scientists directed their attention to these geomagnetic effects.

As cosmic ray studies progressed, it was found that the energy of individual particles, measured in electron volts (eV) was often very high. At room temperature the typical energy of an air molecule is a few hundredths of an electron volt, whereas many cosmic rays have energies well above 1 billion (10^9) eV, and the highest energy yet measured is just over 10^{20} eV. The processes in which such truly astronomical energies are produced have not been identified with any certainty, and theorists continue to grapple with this major puzzle.

When high-energy cosmic ray particles enter the Earth's upper atmosphere, they collide with atoms of the air. Heavy cosmic ray particles, such as nuclei of helium, carbon, or other atoms, are generally fragmented in these collisions, and particles can lose significant fractions of their energy. Emerging from these collisions are streams of particles; some are fragments of the target nuclei and the cosmic projectiles, and others are unfamiliar particles now called mesons. These are not constituents of the matter that we normally encounter; they are radioactive, and their half-lives are short, usually less than a millionth of a second. When they decay or collide with air atoms, more particles, including electrons, are produced. The electrons can radiate photons, which produce yet more electrons, so that cascades develop as this process is repeated over many generations down through the atmosphere, until showers of billions of particles (mostly electrons) can be detected simultaneously at ground level. These extensive air showers provide the only means of detecting cosmic ray particles of the highest energies.

There was a period of intense activity when cosmic ray collisions were studied for their own interest. Because the colliding particles were far more energetic than any that could be produced by cyclotrons, the highest-energy nuclear collisions could be studied only in this way. Many new short-lived particles were discovered during this period, which lasted from the 1930s until the start-up of the Cosmotron at Brookhaven National Laboratory in 1953 and the Bevatron at the University of California at Berkeley in 1955. Since then particle physics has become essentially a laboratory science, almost industrial in its scale, but with the highest-energy particles still the sole property of cosmic ray researchers.

In deciphering the physics of high-energy collisions and fundamental particles, theorists have realized that these particles must have played a critical role during the earliest stages of the explosive "big bang" from which our universe is thought to have started, about 15 billion years ago. In the tiniest fraction of a second, particles and radiation must have interacted in that hot and expanding turmoil. Calculations of the subsequent stages can be tested against present-day observations. The result of one such calculation is the amount of helium produced from the fusion of hydrogen in the early big bang. Helium is continually being produced in nuclear reactions in stars, but not in sufficient quantities to account for all that we observe; production in the big bang seems to solve this problem. The calculation of helium production is very sensitive to assumptions regarding the properties of some elementary particles. Although present-day cosmic ray particles are not directly involved in such calculations, there is no doubt that cosmic ray research paved the way for the modern study of elementary particles.

In a later chapter I will recall this fascinating period in cosmic ray history, but first a more basic question must be addressed: How do we detect and identify cosmic rays, and what is the astrophysical picture we can draw from our findings?

2

AFTER THE ACCIDENTAL DISCOVERY of the leakage of electric charge from electroscopes, it took only a few years before researchers generally agreed that this leakage was indeed caused by a penetrating radiation, of cosmic origin, quite different from X-rays and radioactivity. Radioactive atoms had been found to produce three different types of radiation, named alpha, beta, and gamma by Ernest Rutherford in 1903. The initial classification was based on the degree to which the radiation penetrated lead (which was widely used for shielding) and on the effects seen when magnets were brought near. Rutherford, among others, showed that alpha and beta radiations consisted of particles; alpha particles are identical to the nuclei of helium atoms, and beta particles are the same as electrons and cathode rays. Gamma rays seemed to be more like penetrating X-rays than particles. The newly discovered cosmic rays, however, appeared to have properties that differed from any of these. It was many years before the nature of the cosmic rays (CR) came to be understood, because there are so many different types and their relative proportions change in a complex way as they penetrate the atmosphere.

The primary cosmic rays, both solar CR and galactic CR, that approach the Earth from outside the atmosphere collide with atoms in the stratosphere and produce secondary CR, which may, in turn, have their own collisions. Such secondary and later generation CR were responsible for the electroscope leakage. Most primary CR are protons, the nuclei of hydrogen atoms, which is not surprising if one considers that hydrogen is by far the most abundant element in the universe. On Earth, there is little free hydrogen in the atmosphere but a lot is contained in the molecules of ocean water; hydrogen constitutes about 94 percent of the atoms in the sun and is the element from which other (heavier) atoms form during fusion reactions at high temperatures. Some fusions occurred very early in the development of the universe, in the expansion after the big bang, but others are still taking place, in the interiors of stars and during stellar explosions. The formation of complex nuclei in fusion reactions is called nucleosynthesis.

After hydrogen, helium is the next most abundant element in the sun and stars, and also in cosmic radiation. Among cosmic rays, helium is nearly ten times less abundant than hydrogen. The CR helium nuclei are often called alpha particles, in a continuation of the older nomenclature. In the solar system, all atoms other than hydrogen and helium together amount to only about 0.1 percent of the hydrogen abundance. From this variety of atoms we ourselves are made; indeed, our very existence is excellent testimony to the process of nucleosynthesis.

The primary CR beam also contains small numbers of electrons (less than 2 percent) and gamma rays. Classification of the gamma rays with the "particles" is perhaps a matter of taste, for gamma rays are the highest-energy form of electromagnetic (e-m) radiation. The spectrum of the e-m radiation extends smoothly from the longest-wavelength radio waves to the shortest-wavelength gamma rays (the intermediate visible wavelengths provide the observational base for classical astronomy). The energy in e-m waves frequently behaves as though it is concentrated in bundles (or photons) rather than being spread out, and the photons behave in many ways like particles. Historically, the debate over the nature of light (wave or particle?) has drifted back and forth. Today we use both descriptions, depending on the phenomenon being described.

The main difference between "particle" cosmic rays and gamma rays is that gamma rays have neither mass nor electric charge. Protons, alpha particles, and heavier nuclei have positive charges, and each electron has a negative charge. Gamma ray photons may be the only zero-mass CR "particles." For many years, it was thought that neutrinos, released in many nuclear reactions such as those taking place in the sun and in some radioactive decays, had zero mass. However, in early 1998 there was a report of observations that, it was claimed, indicated that neutrinos had finite though extremely small masses. If this report is sustained by further observations, the implications are very wide-ranging (see Chapter 11). Whatever their true nature, neutrinos continually flood the Earth, passing through it (and us) at all times. Their detection is so difficult that although their existence was first hypothesized in the early 1930s, they were not detected until 1953.

Collisions of nuclear cosmic rays in the atmosphere produce mesons* (mostly pions, previously called π-mesons), which in turn produce muons (originally called mesotrons, later μ-mesons). Charged pions (some positively, others negatively charged) turn into muons and neutrinos in much less than a millionth of a second. Not quite as rapidly, muons turn into electrons and more neutrinos. Neutral (uncharged) pions turn—in less than 10^{-16} of a second—into two gamma rays, which in turn produce electrons, which in the atmosphere produce more gamma rays, and so forth. This ever-changing cascade of electrons, neutrinos, and gamma rays, intertwined with muons, moves through the atmosphere at almost the speed of light. Researchers were not able to interpret sea-level and low-altitude CR observations correctly until they were able to carry out measurements at very high altitudes, where the primary CR alone could be detected. This development, together with the

---

* The naming of these particles was discussed at the Symposium on Cosmic Rays held at the University of Chicago in June 1939. The papers presented there were published in the *Reviews of Modern Physics* 11 (July–October 1939). In a foreword, Arthur Compton noted that "an editorial problem has arisen with regard to the designation of the particle of mass intermediate between the electron and the proton. In the original papers and discussion, no less than six different names were used. A vote indicated about equal choice between meson and mesotron with no considerable support for mesoton, barytron, yukon or heavy electron. Except where the authors have indicated a distinct preference to the contrary, we have chosen to use the term mesotron."

discovery of pions during the period 1946–1950, finally enabled researchers to understand the nature of secondary CR effects at different altitudes.

It is easy today to sit down and write about all these developments, but many of the early results were confusing, and clarity came slowly. The exploratory phase lasted for nearly 50 years. During that time, experimental tools slowly improved, and as a result the complexity of the processes was gradually recognized. Even a brief description of the experimental methods shows how varied and ingenious were the approaches and devices. It is no mean trick to measure the mass, electric charge, and speed of a particle that weighs less than $1/100,000,000,000,000,000,000,000$ (10^{-23}) of a gram and zips through apparatus in less than a hundred millionth (10^{-8}) of a second. Our handle on each particle is its electric charge, but this, too, is tiny. It takes a stream of 10^{19} electrons per second to make up a current of 1 amp, or ampere (a typical electric toaster uses 10 amps). The arrival of one CR particle each second constitutes a current of around 10^{-19} amps. How can such incredibly small particles be detected and identified?

Charged particles can be detected through the ionization they produce. Ionization involves knocking one or more electrons out of an electrically neutral atom, so that the residue, called an "ion," has a net electric charge. As noted earlier, charged particles collide with the nuclei and electrons in the materials they penetrate and knock some electrons off their parent atoms. These electrons can be collected and constitute a minute and measurable electric current. Some electrons move fast enough to produce secondary electrons or other detectable effects. And some CR particles may collide with atomic nuclei. The CR particle may then be fragmented, or the target nucleus may have bits thrown out that will also ionize. These nuclear collisions have been studied for their own interest (to reveal the properties of the colliding particles), and they can also be used to measure the energies and identities of the emerging particles. For the uncharged cosmic rays (gamma rays and neutrinos), detection rests on effects produced when they convert to charged particles or eject charged particles in collisions.

The rate at which a particle loses energy through ionization depends on its electric charge *(Z)* and speed *(v)*. Ionization depends mainly on the ratio Z^2/v^2, which shows that highly charged particles ionize much more than lighter, less heavily charged particles. For instance, an alpha particle (with $Z = 2$) produces 4 times the ionization of a proton ($Z = 1$) with the same speed, and a carbon nucleus (having $Z = 6$) produces 36 times as much ionization. This makes it easier to detect heavier particles, even though they are much rarer. Speed dependence indicates that slow particles ionize much more heavily than do fast particles: the slower the particle, the more time it spends near each atom in detectors, and thus the more of its energy it loses. For particles with speeds above about 0.8 times the speed of light, ionization energy loss no longer depends simply on $1/v^2$, and this is discussed in Chapter 5.

CR particles have kinetic energy (energy associated with motion), which is the source of the energy needed for ionization with corresponding reduction in the particle's kinetic energy. (There are other, nonionizing, ways in which a particle can lose energy.) Ultimately, a particle slows down, loses all its energy, and comes to rest. The stopping distance is the range, a measure of the original kinetic energy. Thus if we can measure the range of a particle, we can infer its initial energy by using calibrated range-energy tables. Kinetic energy depends on both the mass of the particle and its speed. If we can determine the energy (by measuring the range) and the speed (from the ionization), then we can identify the particle; that is, we can figure out its electric charge and mass. Other identification methods have to be used for particles that are too fast to stop within the detectors. In general, to identify a particle we need two different measurements that depend in different ways on mass, charge, and speed.

There are two broad classes of particle detectors. In one, the actual path (or track) of the particle is made visible, and the range may be found directly. Alternatively or additionally, ionization can be measured along the track and changes detected as the particle slows down. In the other type of detector, individual tracks are not seen but the particle's ionization can be measured, either directly or indirectly; the path of each particle is then determined by noting its passage through several detectors in sequence, with the use of very fast electronic timing.

The discovery of cosmic rays was based on ionization in an electroscope. Pioneering experiments, carried out by Millikan before 1920 and by Arthur Compton during the 1930s, also used ionization chambers. In all these studies, the total ionization in a closed container was monitored. This method records the total flow of cosmic rays through the chamber, and thus is a simple way of surveying the variations of CR intensity with altitude (as was done by Millikan) or variations around the Earth, especially at different latitudes (as measured by Compton).

Compton's ionization chamber, shown in Figure 2.1, was 1 foot in diameter and was shielded against local radioactivity by layers of lead. In the center, a spherical container filled with the rare gas argon held a probe connected to a high-voltage source. Electrical leakage between the probe and the container's body was monitored to detect ionization produced by cosmic rays passing through the argon. Although such ionization chambers have largely been replaced by other devices, they played a key role in early discoveries. They were soon supplemented by track detectors and later by complicated electronic systems.

The earliest electrical detector for single particles was devised around 1911 by Hans Geiger, an assistant of Rutherford. In 1929 Geiger and his student Walther Müller introduced an improved version. The design is simple: A thin wire is fixed along the central axis of a metal cylinder but is insulated from direct contact with the cylinder. The cylinder is filled with a low-pressure gas, often argon. A high-voltage battery is then connected between the wire and the cylinder. Because of the gas's low conductivity, only a minute current normally flows around the circuit. But penetration by a charged particle produces ionization in the gas; ions and electrons are accelerated to the wire and the cylinder by the high voltage. Along the way, more ions are produced by electron collisions. An avalanche of ions and electrons arriving at the wire and cylinder constitutes a brief but measurable electrical current, and in this way the passage of the particle can be noted.

Developed at about the same time as the Geiger counter, the Wilson cloud chamber was for many years the most widely used track detector. C. T. R. Wilson, a student of J. J. Thomson at Cambridge University and one of the discoverers of electroscope leakage, was studying the condensation of water vapor. He realized that rapid expansion of gas would produce cooling, as when air escapes from a tire. Vapor condenses when it cools sufficiently, and Wilson found that the vapor condensed on ions and small dust particles. (Legend has it that Wilson's interest in cloud physics came from observing the mists that are common in his native Scotland.) By 1911 he had devised an "expansion

Figure 2.1. The ionization chamber, used by A. H. Compton for his worldwide surveys. The central chamber was heavily shielded by layers of bronze and lead. (From *Physical Review* 43 [1933]: 390)

chamber" (now usually called the Wilson cloud chamber) to study condensation under controlled conditions of expansion and cooling. He showed that X-rays and radioactive particles produced many ions that served as centers for vapor condensation, forming tracks of water droplets that could easily be seen.

Cloud chambers were used extensively in CR and nuclear physics, and major improvements in technique were implemented during the 1930s and the following 20 years. One of the most important findings from use of this device came in 1932, when Carl Anderson and Seth Neddermeyer of Cal Tech discovered the positron, the first example of a particle of antimatter (Figure 2.2). The possible existence of such particles had been predicted by P. A. M. Dirac of Cambridge University in 1929, in his abstract quantum theory calculations. A few years later, Anderson and Neddermeyer also found the first particle whose mass was intermediate between those of the electron and the proton, now known as a muon. Anderson shared the 1936 Nobel Prize for Physics with Victor Hess, belatedly honored for his discovery of cosmic rays. Dirac, along with Erwin Schrödinger and Werner Heisenberg (co-inventors of quantum mechanics), received the 1932 award.

Because tracks in cloud chambers dissipate rapidly, the only useful tracks are those that form during a brief interval around the expansion. In many early experiments, cloud chambers were triggered randomly; it was a lucky accident when a cosmic ray happened to pass through the chamber during the sensitive time and thus appeared sharp in a photograph. An innovation by Walter Bothe changed this situation. Starting in 1924, Bothe and Geiger used two of Geiger's counters for the simultaneous detection of electrons and X-rays in scattering experiments, and Werner Kohlhörster and Bothe then adapted the method to detect cosmic rays. Bruno Rossi, in Florence, greatly increased the sensitivity of this method by developing a new electrical circuit to analyze the signals from a CR telescope—two or three Geiger counters set in a row (Figure 2.3). If a single cosmic ray went directly through the counters, each would give a signal and a *coincidence* would be registered. If the counters were triggered by unrelated particles, their arrival times and electrical signals would not in general be close together.

G. P. S. (Beppo) Occhialini had worked with Rossi in Italy, and when he

Figure 2.2. Cloud chamber track of the positron, discovered by Carl Anderson. Use of a strong magnetic field caused charged particles to follow curved paths. The positron entered the chamber at the lower left and traveled up, through the lead plate across the middle of the chamber. In traversing the lead, the positron lost energy so that its track was more strongly curved when it traveled farther up. Curvature measurements showed that the particle had an energy of 63 MeV (million electron volts) before entering the lead and 23 MeV afterward. (From *Physical Review,* 43 [1933]: 491; courtesy of the Archives, California Institute of Technology.)

joined P. M. S. Blackett in the Cavendish Laboratory at Cambridge University, he applied the coincidence method to Blackett's cloud chamber. Geiger-Müller counters were placed directly above and below the chamber, which was expanded and photographed only if a coincidence was recorded between the counters, indicating the passage of a particle through the chamber. (Occhialini is memorialized in an X-ray astronomy satellite, launched in 1996, that carries the name BeppoSAX.)

Predating both Geiger counters and cloud chambers were photographic materials, which provided the earliest track detectors. Photography was invented in the mid-nineteenth century, long before cosmic rays were discovered or the nature of atoms and electrons was known. When light falls on a photographic emulsion (a thin layer of gelatin containing crystals of silver bromide or silver chloride), it produces submicroscopic changes that show up after suitable chemical treatment (development). Tiny silver grains are produced where the light has been absorbed. When fast charged particles pass through a photographic emulsion, they produce similar changes, and the track of the particle will show up (after development) as a trail of tiny silver grains (Figure 2.4). "Fast" in this context really does mean fast; an alpha parti-

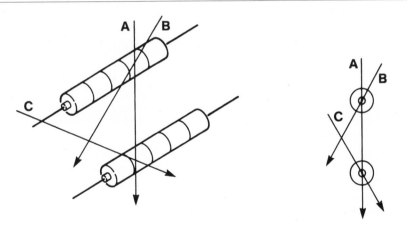

Figure 2.3. The Geiger-Müller telescope. A coincidence is registered only when a single particle (A) passes through both counters. It is unlikely that separate particles (B and C) will go through the counters within a very short interval.

Figure 2.4. Left: Tracks of slow electrons showing the large amounts of multiple scattering that result from the electrons' small mass. Note also the increase in ionization as the electrons slow down. The tracks shown are about 0.2 mm long. Right: Proton tracks (about 0.2 mm long) show a generally much higher degree of ionization and are much straighter, even to the very end where they stop (at the top of the photo). (Photographs courtesy of Peter Fowler, University of Bristol.)

cle whose speed is less than about 1 percent of the speed of light will produce a track consisting of only 1 grain.

Heavy particles with large electric charges leave thick tracks, made up of thousands of grains. Lighter particles with smaller charges produce thin tracks, often with the grains well separated. The particle's speed also affects the shape of the track. Low-mass and slow particles are buffeted around, or scattered, by collisions with the atoms in the emulsion, because of electric forces between each particle and the nuclei in the emulsion atoms; the resulting tracks are full of wiggles. Fast and heavy particles, having greater momentum (mass times speed) leave much straighter tracks (for the same reason that a fast and heavy football player is not easy to deflect). The "straightness" of a track can be put into quantitative terms and reveals the momentum of the particle. Microscopes with a typical magnification of 1,000✕ are used for making track measurements, which is done either by counting grains or by measuring the scattering along a track. The grains one can see in the photo-micrographs are around 1/1,000 mm (about 1/25,000 of an inch) in diameter.

Photographic plates, measuring a few inches on a side, were used by Ernest Rutherford in his early experiments on alpha particles. With these plates he could count alpha particles coming from uranium and thorium, but the emulsions were too thin for long tracks to be seen. The longest-range track of an alpha particle from a radioactive nucleus is about 50 microns (1/20 mm, or 1/500 of an inch). Great improvements in emulsion sensitivity were achieved after World War II.

One application of the photographic method, largely in the biological sciences, has been autoradiography. Tissue samples with radioactive isotope tracers are either held against coated films or have liquid photographic emulsion poured over them. Development, after a period of time that depends on the isotope half-life, shows tracks from the tracers that reveal where in the tissue the isotopes have been transported by biological processes.

The early use of photographic plates in balloon flights to study cosmic rays set a pattern that was followed for many years. In November 1935, Captain Albert W. Stevens and Captain Orvil J. Anderson of the U.S. Army carried special Eastman Kodak plates into the stratosphere aboard Explorer II when they set the altitude record for manned balloon flights at 72,395 feet. Those photographic plates, developed and examined by T. R. Wilkins of the University of Rochester, clearly showed CR tracks, probably the first record of primary CR tracks.

Two Viennese scientists, Marietta Blau and Herta Wambacher, were the first, in 1937, to observe "stars"—tracks in a photographic plate radiating from a common center and produced by fragments of an emulsion nucleus blown apart by a CR collision. The emulsions were not very sensitive, recording only low-energy tracks and heavily ionizing particles. These stars had short black tracks, mostly made by slow protons and alpha particles produced by secondary CR at the mountain altitudes where the plates had been kept for weeks.

Because high-altitude balloons, which became a regular feature of CR work, were not readily available until the late 1940s, most CR experiments were carried out on high mountains. Cecil Powell who had been a student of Ernest Rutherford and C. T. R. Wilson, of the University of Bristol, had been one of the early researchers who used photographic emulsions for CR detection. After World War II, Powell was joined by Occhialini, and they worked with chemists at Kodak and Ilford, encouraging them to produce more sensitive emulsions. Some of these were taken for exposure to cosmic rays at the research stations above 10,000 feet, on the Pic du Midi in the French Pyrenees and the Jungfraujoch in Switzerland (Figure 2.5).

In 1947, in the improved emulsions exposed to cosmic rays on the Pic du Midi, Powell, Occhialini, and Giulio Lattes discovered pion decay, identified by the structure of the pions' tracks and those of their characteristic daughter particles, the muons. The existence of mesons had been predicted by the Japanese theorist Hideki Yukawa in 1935. Only after their discovery in cosmic rays were mesons found in the laboratory, at the new large cyclotron of the University of California at Berkeley.

For many years this area of research provided a steady flow of CR and particle discoveries, first from mountain-altitude plates and cloud chambers and later from photographic emulsions taken to the stratosphere by plastic balloons. An attractive feature of photographic emulsions was their relative cheapness and the simplicity of the instrumentation—no small matters in the depressed conditions of postwar science. Emulsion groups quickly sprang up in London, Dublin, Paris, Brussels, Milan, Genoa, Padua, Rome, Copenhagen, and Bombay. In the United States, early groups were located at the Universities of Rochester and Minnesota. Another group, headed by Maury Shapiro, a former student of Compton, was at the Naval Research Laboratory in Washington, D.C. Emulsions cost around a dollar per cubic centimeter; sensitivity improved and emulsions as large as about 50 cm (20 inches) square and up to 1 mm thick were produced. These required great care in development, and special microscope lenses had to be produced, but the overall requirements were still far more modest than what is now the norm for a satellite-borne experiment.

During the same period (1945–1955), technical advances were transforming the cloud chamber. With careful design the chamber size was enlarged to several feet. Some chambers were optimized for ionization measurements,

Figure 2.5. Left: The Pic du Midi Observatory, in the French Pyrenees. At the high-altitude research station, cloud chambers were operated by groups from Manchester University and the Ecole Polytechnique, and some of the early emulsions were exposed. Right: Cosmic ray physicists on the Pic du Midi, 1949. From the left: Beppo Occhialini (University of Brussels); from the right: Jean Rosch, director of the observatory, Max Cosyns (University of Brussels), P. M. S. Blackett (Cambridge University). (Photographs courtesy of George Rochester.)

and stronger magnetic fields were used for particle deflection. Other chambers incorporated parallel metal plates in which particles could interact. In some arrangements two chambers were operated, one with a magnetic field above a multiplate chamber, so that researchers could identify particles in the top and then examine their later collisions in the chamber below. Like emulsions, the cloud chambers were widely adopted, with major laboratories at Cal Tech, MIT (the Massachusetts Institute of Technology), Indiana University, the Ecole Polytechnique, and Manchester University, among others.

Track detectors require researchers to look at photographic emulsions through a microscope or examine photographs of cloud chambers, and make decisions on the basis of observed tracks. These are tedious procedures that

Figure 2.6. Bristol University balloon, 1953. The inflated balloon is being wheeled out to the tarmac for launch. Inflation took place inside a 120-feet-tall hangar, originally used for airships. Hydrogen was available through a 10-inch spigot on the hangar floor. (Photograph: by the author.)

depend heavily on the perceptiveness of "scanners," laboratory technicians who are skilled at recognizing unusual tracks. Scanning a photographic emulsion is laborious, comparable to reading the Sunday *New York Times* with a hand lens that can show only three or four letters at a time. With the newer electronic nonimaging techniques, however, much of the labor takes place during construction of the apparatus, before any particles are detected. Complex arrays of counters can detect particles automatically and thereby accumulate enormous numbers of them, with rapid analysis following through the use of computers. In the old days, with a typical cloud chamber or photographic emulsion measurement of cosmic rays, thousands of tracks were scanned to accumulate a few hundred special events. Papers reporting the discovery of new subatomic particles were often built around single "events." In their papers in 1948, reporting the discovery of very heavy primary cosmic rays, groups from the Universities of Minnesota and Rochester listed a total of 44 particles, with uncertainties in the charge identifications of about 6–8 charge units.

In contrast, an electronic experiment on the satellite *HEAO-3* detected more than 7 million cosmic ray particles during its year and a half in orbit after its 1979 launch (Figure 2.7). Another experiment, designed to identify the individual isotopes of the iron nuclei, had very demanding requirements for particle acceptance. The University of Chicago's system, aboard the *Ulysses* spacecraft, detected 6,028 iron nuclei during 1990–1995, and another billion other particles went through the detectors and had to be rejected. During its first 4 months of operation, CRIS, the Cosmic Ray Isotope Spectrometer aboard the *Advanced Cosmic Explorer (ACE),* which was launched in August 1997, detected over 10,000 heavy particles and could measure the masses of individual particles so well that individual isotopes could be cleanly identified.

Geiger counters required an electrical circuit, controlled by vacuum tubes (like those used in radios before the invention of transistors); this represented the introduction of electronics to CR research. The electronic revolution, stimulated by World War II, finally transformed particle detection, and today electronic methods of detecting particles by their electrical or light pulses have all but displaced track techniques in CR work. Among the array of

Figure 2.7. Engineers installing a cosmic ray detector in the satellite *HEAO-3*. (Photograph courtesy of Martin H. Israel, Washington University, St. Louis.)

newer devices are scintillation counters, which work on the principle that an ionizing particle produces a brief flash of light (a scintillation) when it goes through a clear material such as lucite. The brightness of this scintillation also depends on the ratio Z^2/v^2. The light can be measured electronically by a photomultiplier tube (PMT) which produces a voltage pulse when the scintillation light falls on the tube's sensitive surface.

There is yet another process in which radiation can be produced by fast particles, and it has been used in many experimental situations. In 1934, Paul Čerenkov, in Russia, discovered that light is emitted when an electron travels faster than light in a transparent material. (This does not contradict the principle of relativity—it is the speed of light in a *vacuum* that cannot be exceeded. In any transparent medium, light travels more slowly than in a vacuum. For example, in glass, light travels with about two thirds of its speed in a vacuum.) The counters Čerenkov devised are similar to scintillation counters in requiring photoelectric detectors but respond only to particles moving faster than a speed that depends on the optical material used. Čerenkov counters are widely used for CR experiments and at particle accelerators. Solid-state detectors, consisting of specially prepared crystals such as silicon or germanium doped with lithium, are now commonly used to measure ionization.

Earlier I mentioned gamma rays as one of the components of cosmic rays, but almost all of this chapter has been about the detection of charged particles. Gamma rays present different problems—partly semantic, partly jurisdictional, partly technical. "Cosmic rays" are usually taken to imply high-speed charged particles, and gamma rays sometimes seem to behave like particles. They are often produced in processes involving cosmic rays, so it is reasonable to consider them here. On the other hand, gamma rays are also electromagnetic waves, as are radio waves, infrared, visible light, ultraviolet, and X-rays. A review of all electromagnetic radiation would have to include all of astronomy. Where, along this seamless spectrum, should the line be drawn? For the purposes of this book, a convenient division can be made to include gamma rays with energies above half a million electron volts, because these high-energy photons are often produced by or along with cosmic rays.

The detection of gamma rays draws on many of the techniques developed for particle detection. Gamma rays can produce ionization, and different types of ionization detectors can be used, depending on the gamma ray energy. High-energy gamma rays (above about a million electron volts) can convert into pairs of electrons and positrons. These particles, in turn, can be detected in ionization detectors or produce tracks in a spark chamber. In this device, thin parallel metallic sheets are connected to high positive and negative voltages. When a fast particle passes through the sheets, it creates ionization that causes a spark to jump between adjacent sheets, showing where the particle has traveled; successive sparks mark out the particle's track. Like a cloud chamber, a spark chamber must be photographed from two angles simultaneously, to generate stereoscopic views from which a three-dimensional picture of the sparks can be reconstructed by using a computer. The midline between tracks of an electron and a positron in a pair indicates the direction of the original gamma ray. Like starlight, gamma rays travel along essentially straight paths, so each arrival direction is unaffected by the Earth's magnetic field. The angle between the two tracks is a measure of gamma ray energy; smaller angles correspond to higher energies.

Extremely high-energy gamma rays (above about 10^{15} eV) can produce cascades of electrons, positrons, and secondary gamma rays in the atmosphere, as do very high energy protons and nuclei. By the time this cascade reaches ground level, a million or more particles can arrive almost simultaneously, spread out over a large area. These extensive air showers are the only effective means of detecting the highest-energy gamma rays and also the highest-energy charged particles, which produce similar showers. I explore them in greater detail in Chapters 6, 7, and 9; for now the important point is that the sea-level detection of showers uses ionization and scintillation counters similar to those described earlier.

The detection of cosmic rays, as we have seen, has required the invention of a whole arsenal of new techniques. Over the years the detectors have grown from small photographic plates and electroscopes to giant electronic systems weighing tons—but still capable of being carried on spacecraft. Accompanying this increase in size, complexity, and cost of the hardware has come the need for laboratories to combine their efforts and share their strengths. These cosmic ray groups are still far smaller than the gigantic collaborations that are common in elementary particle physics, but it is perhaps the only way of investigating many of the questions at the forefront of modern astrophysics.

In the following chapters, I look at the results obtained from this exploration and the inferences that can be drawn about the nature and origin of cosmic rays.

3

APART FROM AIRPLANE FLIGHTS, we live our lives on the Earth's surface, drawing comfort from its seeming solidity. It should be no surprise, therefore, that the earliest thinkers considered the Earth the center of the universe, with the sun, planets, and stars traveling sedately around us in their divinely assigned orbits. Only relatively recently, on the scale of human history, did we recognize that we, too, are moving, with the Earth circling the sun. The entire solar system orbits around the center of our Milky Way galaxy, and, on an even larger scale, the galaxy participates in the general expansion of the universe.

To reach this understanding we have relied on astronomical observations, made first with our unaided eyes and then with telescopes. In all these observations we have depended on an important fact: with minor exceptions, the light we see (such as that from the stars) has come to us along straight-line paths. That is, we generally do not look in one direction to see light coming from other directions. With cosmic rays, however, the situation is almost completely reversed.

Observing cosmic rays from the Earth's surface is like making astronomical observations from the bottom of an ocean. In such a situation, the constant swirling of the ocean surface would distort all images and make them quiver. In addition, the observer would have to correct for the bending (refraction) of light as it passed from the air into the ocean to determine the true direction of an astronomical object. An occasional oil spill would change the amount of light absorbed, as would changes in the quantity of plankton, and the degree of this absorption would vary with the light's wavelength (color). The resulting pictures constructed of the heavens would certainly be distorted.

Cosmic rays detected in apparatus at the Earth's surface or even in the stratosphere are either survivors from primary CR or offspring of those survivors. They have survived collisions with gas atoms in interstellar or interplanetary space, and they also may have collided in the atmosphere. Raw observations thus provide only the first stage in constructing a picture of the CR beam and its journey from distant sources to the Earth.

Just as important as the effects of collisions is the influence of the Earth's magnetic field. In many ways the Earth behaves as though it had a strong bar magnet buried deep within it, about 200 km from its center and not quite aligned with the axis of the Earth's rotation. The extensions to this imaginary magnet come to the Earth's surface at one geomagnetic pole near 79°N, 69°W, north of Thule in Greenland, and the other close to 79°S, 110°E, in the Australian Antarctic. Worldwide surveys of this field have led to the mapping of a grid of imaginary lines of magnetic latitude and longitude, similar to the geographic parallels of latitude and meridians of longitude.

What is meant by a magnetic field? The term "field" denotes a region of space where a force (in this case, magnetic) can be detected and where the strength of that force can be specified at all points. The field strength may be the same throughout a region or it may vary greatly, depending on the nature of its source. It is not necessary for anything visible or tangible to support the field: the force of a simple magnet can be felt when we try to pull an iron object away from it.

Electromagnetic waves carry programs from transmitters to radio and television antennas. All e-m waves travel with the speed of light, carried by electric and magnetic fields whose strength at every place oscillates very rapidly. If we could momentarily freeze a wave, we would find that it forms an endlessly repeating pattern along which places of maximum field strength are regularly spaced one wavelength apart. The fields oscillate many times each second with a well-defined frequency. The general formula for all waves is frquency times wavelength equals speed. In television transmission the fields change with frequencies of around 100 million oscillations per second (100 million hertz or 1 MHz), and the wavelengths are around 300 cm. FM radio uses similar frequencies, but AM radio frequencies are lower, around 600,000 hertz (600 kHz). The visible part of the e-m spectrum has wavelengths extending from about 4×10^{-5} cm at the blue end to 7×10^{-5} cm in the deep red. Much shorter wavelengths include ultraviolet rays, X-rays, and gamma rays, while the longer wavelengths extend into the infrared and then radio.

The Earth's magnetic field does not oscillate rapidly as an e-m wave does but changes slowly over tens of thousands of years. It extends far into space, with steadily decreasing strength; the field drops by half at a distance of 1,000 km out from the Earth's surface. Farther out, the field continues to decrease until it encounters the ionized gas that is continually spreading out from the sun. This ionized gas, called the solar wind, was discovered as recently as the early 1960s, although its presence had been suspected many years earlier.

The interaction between the solar wind and the Earth's magnetic field is now studied with satellites and long-range space probes, but a starting point is an understanding of the basic phenomenon of magnetism. The behavior of magnetic rocks, called lodestones, has been known since ancient times. These have been used for navigation since at least the twelfth century, and simple magnetic compasses, such as the kind one attaches to a dashboard, are still in use. Magnetic compasses do not point accurately to true north, however; one must compensate for local effects as well as for the displacement of the north magnetic pole from the geographic pole.

The first systematic study of magnetism was published in 1600 by William Gilbert, physician to Queen Elizabeth. This work had a strong influence on Johannes Kepler, who developed his theories of planetary motion under the mistaken impression that a magnetic force from the sun drove the planets in their orbits. Kepler made other incorrect assumptions that eventually canceled each other out in his calculations; as a result he did, in fact, deduce correct laws of planetary motion. But it was Isaac Newton, about 60 years later, who correctly identified gravity as the controlling force.

Edmond Halley, so well remembered for his association with the famous comet, was also renowned for his magnetic survey. He carried out this work during three voyages between 1698 and 1702 for the British Admiralty, which had a continuing interest in the improvement of navigational aids. In what was probably the first scientific expedition completely supplied and funded by a government, the Royal Navy provided Halley with a ship and all the instrumentation needed for his survey. Royal recognition came in 1701, on the command of King William to the Navy Board: "In obedience to His Majesty's Commands . . . desire and direct to be paid unto Captain Ed. Halley . . . the Sum of two Hundred Pounds . . . in consideration of his great Paines and care in the late Voyage he made for the discovering the Variation of the Needle" (N. J. Thrower, *The Three Voyages of Edmond Halley,* p. 321). (The "variation" is the angle between the true north and the direction shown by the magnetic compass needle.)

The relevance of magnetism to cosmic rays is that a stream of electrically charged particles (such as CR) constitutes an electric current. The effects of a magnetic field on an electric current have been known since the pioneering work of Hans Christian Oersted and others in the early nineteenth century— research that led to the invention of electric motors and generators.

The paths followed by cosmic rays as they approach the Earth are influenced by the Earth's magnetic field. Several factors control the path of a CR: its initial direction of travel; its mass, speed, and electric charge; and how the magnetic field varies with distance from the Earth. As a result, the direction of motion changes continually as a CR approaches the Earth, resulting in a highly convoluted path. We can now see why the path of a CR particle detected by a CR "telescope" (composed of Geiger counters, for example) is not a reliable indicator of the particle's direction before it came to Earth. Indeed, because of magnetic fields even in the great distances between planets and the greater distances between the stars, a CR path is never simply a straight line from its source to the Earth (see Figure 3.1). I elaborate on this in Chapters 5 and 8 in the discussion of galactic CR and their history.

Early surveys of the Earth's magnetic field were, of necessity, carried out close to the Earth's surface, but with the advent of the space age, direct measurements have been extended to great distances but still within the solar system. A mathematical model based on these surveys enables us to calculate in detail (with the use of computers) a CR's path through the Earth's magnetic field. The calculation is simple but laborious: you start with a CR at some location far from the Earth and specify its charge, mass, speed, and direction of motion. The magnetic force that deflects the particle depends on a complex combination of these factors. Then, using a computer and a model for the field, you calculate the field strength at that place and compute the force that the particle will experience because of the field. This force changes the particle's direction by a small amount, so you can calculate where the particle will be a short time later. Now repeat the calculation, allowing for the changed direction and also for any difference in field strength. Once more, calculate where the particle will be after another short interval. Then repeat the whole process—millions of times. Subtleties in the programming of such a calcula-

Figure 3.1. Some computed cosmic ray orbits, showing the effect of the Earth's magnetic field on incoming charged particles. The complexity of each orbit depends sensitively on the particle's charge, momentum, and direction of travel. Some particles may spiral around the Earth many times before descending to a low altitude. (Courtesy of Niels Lund, Danish Space Research Institute.)

tion can keep the computing time (and cost) to a minimum while protecting against errors that could accumulate when a final result is based on millions of successive calculations.

In working out the solution of a theoretical physics problem (such as CR paths), you start with a few equations that relate various measurable quantities to one another. With some types of equations, however, there is no exact solution in the form of a general formula. You then have two choices: either produce numerical solutions for specific cases, or go back to the starting equations and see whether you can simplify them and then derive exact solutions. Simplified equations refer to idealized circumstances, but they can serve as a guide in working numerically with the complete equations.

The precise shapes of CR paths pose one such problem, with no general formula as a solution. Fortunately, it has turned out that one can solve equations for a simplified model of a magnetic field similar to the Earth's. This approach was first used by the Norwegian mathematician Carl Störmer starting in 1903, relatively soon after the electron was discovered. As he described it, his inspiration came when "my colleague Kristian Birkeland showed me some of his very beautiful and striking experiments on the movement of cathode rays in magnetic fields" (*The Polar Aurora*, preface). Störmer became interested in polar auroras, the impressive atmospheric displays in which large regions of the upper atmosphere glow with brilliant colors in changing patterns (Figure 3.2). The discovery of electrons and the strong correlation between auroral activity and solar activity (such as sunspots) stimulated speculation that auroras might be generated by electrons coming from the sun. Störmer set out what has become the classic mathematical description of electron orbits in the geomagnetic field. In his research on the aurora, he focused on relatively low-energy electrons; a later extension of the theory to high-energy particles was prompted by the growing interest in the nature and origin of cosmic rays and by measurements of CR intensity made at widely separated locations.

In 1927 and 1928, the Dutch physicist Jacob Clay published the results of his measurements of CR intensity during voyages between Genoa and the Dutch colony of Java. Using ionization chambers, Clay found that CR intensity had its minimum value at the magnetic equator and rose by about 10 percent at northern latitudes. Meanwhile, Bothe and Kolhörster, using Geiger counters, had shown that the sea-level CR consisted of charged particles. They went on to point out that Clay's latitude effect could be understood if the primary CR also consisted of charged particles, and if those particles were deflected above the atmosphere by the Earth's magnetic field. Clay's results were confirmed and extended in a worldwide survey carried out by Arthur Compton of the University of Chicago between 1930 and 1933 (Figures 3.3, 3.4, 3.5). Compton made measurements at sea level and also at mountain altitudes, where the latitude effect was even larger.

The theory of the geomagnetic effect on CR attracted renewed attention, first by Bruno Rossi (then at the University of Florence) in 1930, followed by

Figure 3.2. An aurora. High-speed charged particles are guided by the Earth's magnetic field as they enter the upper atmosphere. In collisions with atmospheric atoms and molecules at typical altitudes of 180 km, energy is transferred from the incoming particles and radiated to produce the aurora. This view of the Aurora Australis was taken by the crew of the *Space Shuttle Endeavor* in April 1994, while in low-Earth orbit. The constellation Orion is clearly visible in the background. (NASA Astronomy Picture of the Day for 20 March, 1999; photograph by STS-59 Crew, NASA.)

the Belgian cosmologist Georges Lemaitre and the Mexican physicist Manuel Vallarta, then at the Massachusetts Institute of Technology (MIT). Their calculations showed clearly that the latitude effect was a consequence of the fact that charged CR particles arrive randomly from all directions before they encounter the Earth's field and begin to be deflected. The calculations of Lemaitre and Vallarta and of Rossi revealed that only the fastest, most energetic particles could penetrate to the region of the magnetic equator. Somewhat less energetic particles would be detectable in the mid-latitudes, and in

Figure 3.3. Victor Hess examining equipment used by Arthur Compton during his worldwide survey. Note the heavy shielding around the equipment, housed on the ship deck. (Photograph courtesy of Washington University Archives, St. Louis.)

the polar regions there would be no restriction on the energies of the detectable primary particles. Overall, the number of particles detected should increase steadily as one moves away from the magnetic equator. The observations of Clay and Compton were thus nicely explained.

The latitude effect showed clearly that primary CR particles had to be electrically charged, but that both positive and negative charges would show the same effect. One prediction from the theory did allow the sign of the charge to be identified: each latitude has a cutoff energy below which no vertically arriving particles will be found. The cutoff energy is different for positive and negative particles arriving in nonvertical directions. Lower-energy particles of positive charge would penetrate if they arrived from the west, but for positive particles arriving from the east a higher energy would be needed. For negative particles, the effect would be reversed. More low-energy particles than high-energy ones are found among the cosmic rays, and thus this east-west effect predicts that, if primary CR are predominantly positively charged, more CR should be detected by a Geiger telescope directed to the west than by one di-

Figure 3.4. The results of Compton's survey, showing cosmic ray intensity at sea level for various geomagnetic latitudes. The four curves represent data taken during different seasons. (Adapted from *Physical Review* 52 [1937]: p. 808.)

rected to the east. Confirmation of the east-west effect came in 1938 from measurements carried out by Thomas H. Johnson of the Bartol Foundation and the Carnegie Institution of Washington. The inference was clear—the majority of primary CR were positively charged.

Improvements in the capacity and speed of computers have enabled researchers to explore the calculation of CR paths in detail unimaginable by Störmer and even by Lemaitre and Vallarta, who worked with an early computer that used electric motors and mechanical linkages. Because a typical CR

Figure 3.5. Compton relaxing during a voyage. (Photograph courtesy of Washington University Archives, St. Louis.)

orbit can loop around the Earth many times (see Figure 3.1), it has to be followed (in the computation) through millions of small steps. An even more complex picture emerges when many CR path calculations are examined. Diagrams of many paths resemble a tangle of spaghetti. Some paths, after several loops, provide unbroken connections to the Earth's surface from regions far beyond the solar system. Some cosmic rays approach the solar system, travel along these paths, and arrive close enough to the Earth to be detected; they display the latitude and east-west effects just described. But it turns out that there is also a quite different family of paths that loop around the Earth, neither dipping down to the surface nor escaping. A particle traveling along such a path would be trapped (like a celestial *Flying Dutchman*) in a permanent orbit around the Earth, looping between the northern and southern hemispheres and drifting around in longitude but never escaping. Cosmic rays approaching from far away cannot connect directly to these paths, and the only way a particle can get into one of these perpetual loops is to be born in it (through some nuclear process) or be bumped into it (by a collision with another particle). For this reason, no importance had been attached to these strange orbits by CR researchers, even though their mathematical existence had been known from the calculations of Störmer, Lemaitre, and Vallarta. Then, totally unexpectedly, these orbits were rediscovered in 1958.

Under the auspices of the International Council of Scientific Unions, a coordinated international study of the Earth, the International Geophysical Year (IGY), was planned for 1957–58. Many types of observations were to be carried out simultaneously all over the world, to reach a better understanding of the Earth and its relation to the sun. The first Earth-orbit satellites were designed to extend these observations into near-Earth regions. On October 4, 1957, we awoke to the news that *Sputnik I* was in orbit; reports carried recordings of its high-pitched beeping signal. The political response in the United States was prompt and intense. Increased federal funds were voted for scientific research, and major efforts were undertaken to modernize the science curricula in our schools.

Sputnik I did not carry CR instruments, but a monitoring of its radio transmissions revealed properties of the ionosphere, the ionized region of the

Earth's upper atmosphere. The extended gravity field of the Earth could be probed by tracking the satellite in its orbit. *Sputnik II,* launched in November 1957, did carry CR detectors, but nothing unusual was reported by Sergei Vernov and his colleagues at the Academy of Sciences in Moscow.

After a few initial failures, the first U.S. satellites were launched in February and March 1958, carrying Geiger counters to monitor cosmic rays. The CR experiment, designed by James Van Allen and his group at the State University of Iowa (Figure 3.6), surprised everyone with its early results; the signals telemetered back to Earth were quite different from what had been expected.

Figure 3.6. James A. Van Allen performing preflight tests on his radiation detectors, which were used in the Jet Propulsion Laboratory spacecraft *Pioneer IV,* launched in March 1959. These detectors, and similar ones on *Pioneer III,* launched in December 1958, provided the first radial scans of the large-scale structures of the inner and outer radiation belts, then only recently discovered. The intensity of the primary cosmic radiation far beyond Earth's magnetic field was also measured. (Photograph courtesy of J. A. Van Allen, University of Iowa.)

During the decade preceding the satellite voyages, rocket explorations had found that CR intensity increased up to about 24 km (as had been already known) but then decreased to a steady value at the highest altitudes reached, around 150 km. The U.S. *Explorer I* and *II* satellites, however, were reaching far greater altitudes. In very eccentric orbits, they looped from their closest approach (perigee) at about 250 km (150 miles) above the surface out to a maximum distance (apogee) of about 2,600 km (1,600 miles) every 2 and 1/2 hours. And what Van Allen noted was that the cosmic ray intensity first increased slowly with altitude, but above about 2,000 km the CR seemed to vanish, as if the counters were not working. Then, when the satellite dropped back below 2,000 km, the counting resumed. The interpretation, confirmed by testing duplicate counters in the laboratory, was that the Geiger counters on board were being overloaded—which meant that the counting rate had to be at least 15,000 times greater than the usual CR rate. This was confirmed by *Sputnik III* and *Explorer IV* in 1958, both by then carrying counters designed to cope with the greater particle arrival rates.

The satellite data also showed that the intensity of this new radiation was correlated with magnetic but not geographic latitude; thus, like cosmic rays, the radiation had to be controlled by the geomagnetic field. Later space probes, some in orbits relatively close to Earth but others traveling out to distances of several Earth radii, provided the data to complete the picture. It soon was seen that the Earth was surrounded by two gigantic radiation belts (Figure 3.7) in which large numbers of high-speed electrons and protons were moving, trapped in orbits that, only 20 years earlier, had seemed to be just a mathematical curiosity.

Why didn't the first two *Sputniks* discover these radiation belts? Although *Sputnik II,* unlike *Sputnik I,* carried counters, the shape of its orbit and telemetry restrictions provide the answer. Data were telemetered back to Earth only when the satellite was over friendly territory, and at those times the satellite was not penetrating the dense parts of the trapped particle region. If the Russians had telemetered their data more continuously, in a form that could be understood by receiving stations elsewhere, we would probably now be referring to the Vernov radiation belts rather than the Van Allen belts. Van

Allen announced his discovery in May 1958, and until then there had been absolutely no hint that the Russians had noted anything unusual in their counting rates.

Before examining this trapped radiation, it is useful to take a closer look at the geomagnetic field to see how it controls the movement of charged parti-

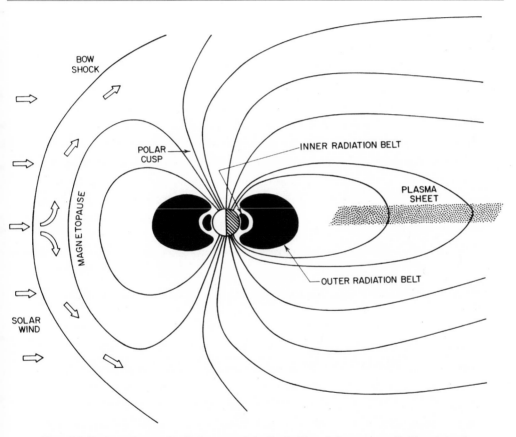

Figure 3.7. Regions of trapped radiation around the Earth. These belts are controlled largely by the Earth's magnetic field, but the influence of the solar wind distorts the outer belt, compressing it on the sunward side and extending it on the side away from the sun. (Courtesy of J. A. Van Allen. Reprinted with permission from James Van Allen, "Radiation Belts," in *Encyclopedia of Physics,* ed. Rita L. Lerner and George L. Trigg, © 1981 Wiley-VCH Publishers, Inc., New York.)

cles. The field around a magnet can be pictured as filled by a family of lines. If a small magnetic compass is used to survey the field of a large magnet, the lines of force will follow the direction of the test-compass at different places (as with a magnet held under a sheet of paper on which iron filings are sprinkled). The term "lines of force" was introduced by Niccolo Cabeo in 1629, but it was Michael Faraday, a scientist of great insight and inventiveness working in London's Royal Institution in the nineteenth century, who developed a quantitative model of magnetic fields—a visualization still of great use and found in most textbooks.

It is also helpful to review the quantities and units that describe the energies of CR particles. The laws governing moving objects and particles were set out in their modern form by Isaac Newton in 1687 in his great treatise the *Principia*, and his methods are still used today. The concept of speed is familiar from everyday experience, but calculating how a particle responds to forces (such as the Earth's magnetic field) requires that the particle's mass be taken into account. Two auxiliary quantities that combine speed and mass are momentum, mv, and kinetic energy, $1/2\ mv^2$, where m denotes the particle's mass and v its speed (or velocity).

These expressions for momentum and energy must be modified when the particle's speed is greater than about a tenth of the speed of light in a vacuum *(c)*, to take into account effects first recognized by Albert Einstein in 1905. Although v must always be less than c, as postulated by Einstein, the particle's mass and kinetic energy can increase without limit. Consequently, it is much more convenient to describe particles by their (kinetic) energies than by their speeds. The usual units for energy are electron volts (eV) and its multiples of one thousand (keV), one million (MeV), and one billion (GeV). (The European GeV is preferred to BeV for 10^9 eV, to avoid confusion. In the United States "billion" denotes one thousand million, whereas in European usage it is one million million.) This terminology dates from the early practice of using high voltages to accelerate cathode rays. The electron volt is an extremely small unit of energy, but quite appropriate to the scale of atomic particles. If we used electron volts for everyday objects, we would find that the kinetic energy of a baseball thrown at a speed of 30 miles per hour could be expressed as

8×10^{19} eV. Yet the highest-energy CR particles have about three times as much kinetic energy, despite weighing 10^{26} times less.

Table 3.1 lists a few representative values for the speeds and kinetic energies of protons (hydrogen nuclei) and electrons that are encountered in CR studies; it can be seen that the energy increases even though the particle's speed approaches its limiting value, c. An alpha particle from the radioactive decay of a radium or uranium nucleus has an energy of around 5 MeV, and CR energies range from tens of MeV to more than 10^{20} eV.

In the Van Allen radiation belts we find electrons with typical energies of a few MeV and protons of tens of MeV up to a few hundred MeV. A particle with an energy in this range will move around a line of force in a spiral path that becomes more tightly coiled as it comes closer to the Earth. Eventually the path becomes momentarily parallel to the Earth's surface, and then the particle starts to move back up. This bounce, or reflection, is governed by the field strength and the particle's momentum; the field acts like a magnetic mirror. After reflection the particle spirals up and loops over to a corresponding mirror point over the opposite hemisphere, where the process is repeated. A trapped particle will bounce back and forth repeatedly between hemispheres while at the same time drifting in longitude. A similar magnetic effect is employed in experiments designed for hydrogen fusion (thermonuclear) reactions: a carefully shaped magnetic "bottle" confines a hot ionized gas (a plasma) so it cannot touch (and vaporize) the walls of its containing vessel.

Particles trapped in the Earth's radiation belts can continue to bounce between hemispheres until some new influence intrudes. Because the magnetic

Table 3.1 Particle speeds expressed as fractions of the speed of light.

Kinetic energy	Electrons	Protons
1 keV	0.063	0.0015
1 MeV	0.94	0.046
100 MeV	0.999987	0.43
1 GeV	0.99999987	0.88

field is not perfectly regular, the drift motion is actually very complex, and some particles can transfer to different field lines that will guide them down into denser regions of the atmosphere, where they will be lost through collisions. Alternatively, particles can be lost by collisions among themselves or by distortions of the magnetic field during times of increased solar activity. Electrons in the inner belt may be lost within a year, those in the outer belt in a few days. The typical residence time of a proton in the inner radiation region is measured in years. Overall, although the details remain to be elucidated, the picture is dynamic, with particles being continually injected, others being lost, and the belts being replenished.

The structure of trapped radiation belts is sensitive to solar influences by way of the solar wind; a typical depiction is shown in Figure 3.7. The inner belt, most heavily populated by protons, has most of its high-energy particles (above 10 MeV) about 1.5 Earth radii (R_E) out, but some of these protons can be found as far as 4 R_E. At its most intense, there are 2×10^5 particles/cm²sec. For lower-energy protons (above 0.1 MeV), the peak of 2×10^8 /cm²sec is around 3 R_E.

The outer belt contains most of the electrons; for those with energies above 1 MeV the peak flux of 3×10^6 particles/cm²sec is around 4.5 R_E, and electrons can still be found more than twice as far out. There are, of course, no sharp edges to these regions; as Figure 3.8 shows, the particles are spread around the Earth in a doughnut-shaped region in which the filling varies in density. In the inner belt, the number of protons peaks at about 15 per cubic meter. A few trapped helium nuclei have also been detected, about one for every 5,000 protons, and a smaller number of oxygen ions with energies up to about 20 keV.

The numbers given here for the populations of the trapped particles are useful as guides, but those populations do vary greatly. The structure of the outer regions of the Earth's field is sensitive to the magnetic field carried by the solar wind, and that field in turn is subject to the variable level of solar activity, which generally follows an 11-year cycle with sporadic additional outbursts. In the region where the solar wind sweeps by the Earth, the collision of the two magnetic fields creates a complex boundary layer. The Earth's magnetosphere is somewhat compressed on the sunward side and is pulled out considerably more on the opposite side of the Earth. The inner radiation belt is little affected by these changes, but the distant regions of the outer belt are much more responsive to them.

The basic correctness of this model of the Earth's field with its trapped par-

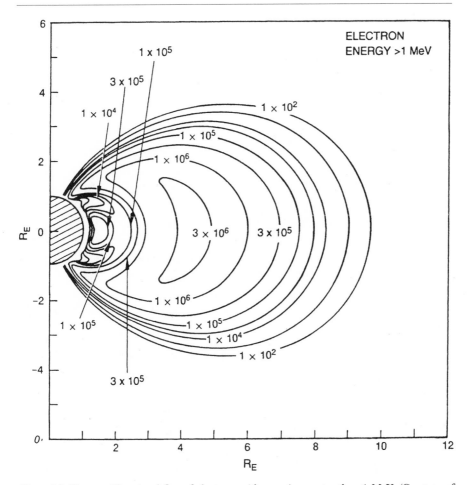

Figure 3.8. The omnidirectional flux of electrons with energies greater than 1 MeV. (Courtesy of Martin Walt; based on data from the National Space Science Data Center.)

ticles has been confirmed by experiments with high-altitude nuclear explosions. These tests were carried out from 1958 to 1962, before growing public concern over radioactive fallout led to the adoption of the first test ban treaty. In the test explosions, large numbers of high-energy electrons were released and their progress followed as they spiraled along the magnetic field lines that passed through the locations of the test shots. In a particularly dramatic demonstration, the Argus shot detonated over the Johnston Islands in the Pacific produced an aurora later seen over Samoa, about 2,000 miles to the south. These two locations are at magnetic conjugate points (symmetrically placed north and south of the magnetic equator), and electrons spiraled around the field lines until they reached a mirror point over Samoa where atmospheric collisions created the aurora, a rare phenomenon at tropical latitudes. A densely populated radiation belt, lasting for some years, was produced by the Starfish test of July 1962. The Soviet Union also had several test explosions.

Another magnetosphere with a population of trapped particles was discovered in 1955, when Jupiter was identified as the source of intense and variable radio signals. These radio emissions, studied from the Earth and from the *Pioneer* and *Voyager* long-range probes, revealed that Jupiter possesses a magnetic field much stronger than the Earth's, with large numbers of trapped electrons and heavy ions. Jupiter is about five times farther from the sun than the Earth is, so solar-wind effects on Jupiter's magnetosphere are different in scale. The phenomenon of radio emission by fast electrons traveling along curved paths in a magnetic field (synchrotron radiation) was first elaborated for astrophysical settings largely by the outstanding Russian theorists Iosef Shklovsky and V. L. Ginzburg. A curiosity of Jupiter's emissions is that they are strongly influenced by Io, the innermost of the four moons discovered by Galileo in 1610. The cause of Io's influence has not yet been found, but it should be noted that Io is well inside Jupiter's trapped electron region, in contrast to our moon, whose orbit places it about ten times farther out than our outer radiation belt.

The inner planets Mercury, Venus, and Mars do not have magnetic fields that are strong enough to support radiation belts, but large numbers of trapped particles have been detected around Saturn, Uranus, and Neptune, which do have strong fields. Planets with strong fields, rapid rotation, and moons within the radiation belts, present situations that are much more complex than that for the Earth. In addition, some moons such as Io have their own atmospheres, and oxygen, sulfur, and sodium ions have been detected in these distant radiation belts.

Our understanding of the sources of the Earth's trapped particles is far from complete. Soon after Van Allen's discovery, a plausible explanation attributed the trapped particles to primary CR collisions in the Earth's atmosphere. In many of these collisions, neutrons are ejected from atomic nuclei. With a half-life of about 11 minutes, each free neutron decays to produce a proton, an electron, and a neutrino. Because neutrons carry no electric charge, they are unaffected by the geomagnetic field and can travel directly away from the sites of their release until they decay. In this way they can reach any point in the magnetosphere and produce electrons and protons in places that charged particles could not otherwise have reached as quickly or at all. Some of these decay electrons and protons will have the right speeds and directions of travel to allow them to be trapped. This mechanism, however, cannot account for all of the trapped particles apart from the most energetic protons (with energies above about 50 MeV) in the inner belt. Recent theory attributes the origin of most particles to a combination of solar wind, CR interactions, and the ionosphere. Through processes that we understand only partially in this complex astrophysical setting, slow particles are accelerated by changing electric and magnetic fields. A similar situation exists for the explanation of the auroral particles.

Geophysical studies have thus provided (in the latitude and east-west effects) the basis for identifying the primary CR as electrically charged particles rather than the (electromagnetic) photons that Millikan had initially preferred. When the IGY monitoring of cosmic rays led to the entirely unanticipated discovery of trapped particles, it was again geomagnetic theory that provided the framework for a model.

4

THE EARTH'S LOCATION in the solar system preserves a delicate thermal balance. If we were much closer to the sun our atmosphere would lose its essential lighter gases, and if we were much farther away life either would have evolved differently to cope with freezing temperatures or perhaps would never have developed at all. The sun's energy, originating from fusion reactions deep within, diffuses to the surface and radiates in all directions, with less than one part in a billion striking the Earth. Because the solar interior is so dense, the energy takes more than a million years to travel from the core to the surface, but then only another 8 minutes to go the additional 150 million km to the Earth. (This distance, equivalent to 93 million miles, is often used as a measure of distances within the solar system and is called the astronomical unit, or AU.)

The sun maintains its radiant output at a generally steady level, a baseline known as the quiet sun. At times an active sun, with an increased number of sunspots, can create on Earth a disruption in long-range radio transmissions. Other indications of increased solar activity will be discussed later in this chapter.

More than electromagnetic energy is departing from the sun. The bright photosphere that we see has a temperature of around 6,000°K, but the overlying corona (spectacularly visible during solar eclipses) has a temperature range that extends to over a million degrees. At this temperature some ionized gas constituents are moving at speeds great enough to escape from the solar gravitational attraction, flowing out through the solar system as a solar wind. The flow amounts to about 300,000 tons each second, but this seemingly enormous quantity still amounts to only one part in 10^{14} of the solar mass each year.

Although it might seem incredible that the sun's loss of hundreds of thousands of tons of ionized gas each second could escape our notice, that was indeed the case until as recently as the 1960s, when the existence of the solar wind was firmly established. The solar wind is now easy to detect with particle detectors carried on satellites well above our atmosphere, but before the space age the only indications of its presence were less direct. There had long been speculation that the sun influenced our weather, and of course astrology was based on the supposition that events on Earth are controlled by the sun and other heavenly bodies merely by their positions. The first scientific intimation of a solar influence came in 1859, when R. C. Carrington, an English amateur astronomer, observed a solar flare, an intense local brightening on the face of the sun close to a sunspot. At the same time, monitors of the Earth's magnetic field recorded changes from their usual values. Such events are termed magnetic disturbances, or storms. The nature of any physical way in which the sun could control these phenomena on Earth was quite unknown. Carrington also noted that about 12 hours after the flare a major magnetic storm accompanied by auroras was seen in both hemispheres. The possible connection between solar activity and auroras was a subject of continued investigation, but the scattered speculations were inconclusive.

A related solar phenomenon also stimulated the search for possible solar-terrestrial connections. On many occasions spots had been observed on the disk of the sun, but they were invariably ascribed to the passage of planetary bodies such as Mercury or Venus in front of the sun, since people were reluctant to challenge the presumed unblemished nature of the sun. After Galileo's innovative application of the telescope to astronomy, in 1609, however, this interpretation had to change. One of his early discoveries was that the spots were truly attached to the sun. Continued observations showed that the spots moved across the sun's face as though the sun itself were rotating; confirmation came when spots were seen to disappear at one limb of the sun and reappear round the opposite limb about 2 weeks later, implying a 27-day rotation period. (The true rotation period is closer to 25 days; the longer apparent period is a result of the fact that the observations are being made from the Earth, which is not at rest but is moving in its orbit around the sun.)

The number of sunspots visible at any one time is highly variable. In 1877 Rudolph Wolf of Zurich compiled records of sightings dating back to 1610 and found that sunspots occurred in cycles, their numbers increasing and decreasing over intervals of about 11 years (Figure 4.1; see also Figure 4.2). The Zurich sunspot number, still our most direct indicator of solar activity, has a periodicity similar to that of geomagnetic disturbances, and it was natural to think that the latter might be related to the sunspot cycle. No physical link could be found, however, and the connection remained speculative. No less a figure than Lord Kelvin argued, in 1892, that "we may be forced to conclude that the supposed connection between magnetic storms and sunspots is unreal, and that the seeming agreement between the periods has been a mere coincidence." A few years later C. A. Young, a Princeton astronomer, expressed a more cautious judgment: "From the data now in our possession, men of great ability and laborious industry draw opposite conclusions" (*The Sun*, pp. 169 and 162).

Additional and more persuasive evidence for the extended influence of the sun accumulated from three rather different lines of research. First, the 27-day periodicity in geomagnetic activity persisted over many years. Over a short period one might be willing to regard the similarity of 27-day cycles as accidental, but not when the effect persists for 30 years and covers nearly 400 cycles, with the average length so similar to the solar rotation period.

Further support came from CR studies, but in an unanticipated way. Soon after the discovery of cosmic rays, a search began for variations in their intensity, not only with altitude but also with time. The early records of CR intensity showed large variations, up to 15 percent, but there was no consistency between different measurements. More careful work, together with improved corrections based on better knowledge of CR processes in the atmosphere,

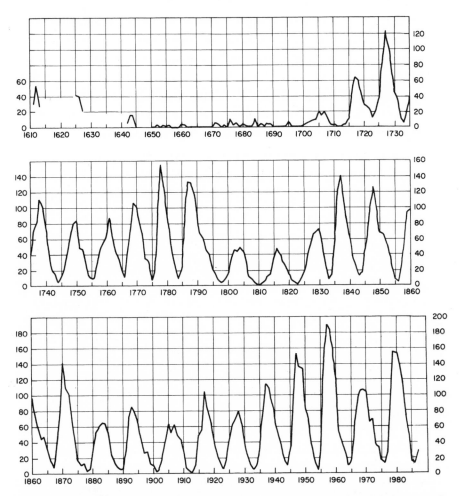

Figure 4.1. Annual mean sunspot numbers from 1610 to 1987. Since about 1710, there has generally been an 11-year periodicity, but with different numbers of sunspots seen in successive maxima. For the years preceding 1700, the 11-year periodicity has not been as firmly established. Sunspot observations were not routine before the 1609 invention of the telescope and were still sparse for many years subsequently. However, the paucity of sunspots during the years 1650 to 1700 is considered well established and this period is known as the Maunder Minimum. It has been correlated with many climatic effects, including low temperatures worldwide, early frosts, increasing glacier sizes, and late blooming of cherry trees in the royal gardens in Kyoto. (Courtesy of John A. Eddy, High Altitude Observatory).

proved that daily variation was itself variable, but on the average less than 0.5 percent. The modern era of CR monitoring started in 1936 when Scott Forbush, of the Carnegie Institution in Washington, set up a network of stations with recording ionization chambers in Greenland, the United States, Peru, and New Zealand. Forbush found that the intensity of cosmic rays reached a maximum each day, usually around 3:00 P.M., and then declined to a minimum around 12 hours later, with the swing from maximum to minimum being only about 0.15 percent on average. By 1957 Forbush had sufficient data to conclude that there was a correlation between long-term CR variations and solar activity, but an inverse one: over the 11-year solar cycle, when sunspot

Figure 4.2. Monthly sunspot numbers during 1953–1999. (National Optical Astronomy Observatories.)

numbers increased, CR intensity decreased (Figure 4.3). Finally, also in 1957, H. V. Neher of the University of Minnesota, who had been conducting high-altitude balloon flights each year near the magnetic pole at Thule, Greenland, discovered that low-energy CR protons were significantly more numerous when solar activity was minimal.

These varied observations led to the surprising conclusion that solar activity, as indicated by the 11-year sunspot cycle, was not producing more cosmic rays but rather was somehow preventing galactic cosmic rays from reaching the Earth. Quite unexpectedly, the last piece in the puzzle came from studies of comet tails.

Comets occupy a distinctive place in astronomical folklore. Their unpredicted appearances used to be taken as omens, and Halley's comet is depicted

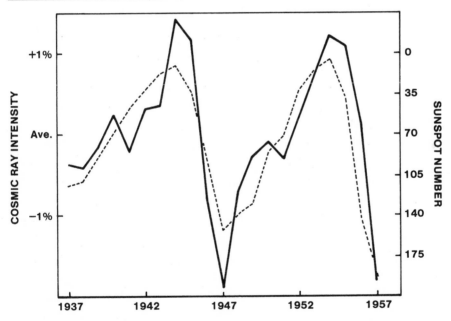

Figure 4.3. Correlation between cosmic ray intensity (solid line, scale at left) and sunspot number (dashed line, scale at right), through two solar cycles. Increased solar activity produces a decrease in the cosmic ray intensity at Earth. (Adapted from a figure by Scott Forbush, *Journal of Geophysical Research* 63 [1958]; 657; © American Geophysical Union.)

in various early works of art such as the 1066 Bayeux tapestry and Giotto's 1304 fresco in the Scrovegni Chapel in Padua. The return of this comet in 1986 showed that these objects can still arouse great interest. The often spectacular comet tails always point away from the sun, so they will trail behind a comet as it approaches the sun and flow ahead of the comet as it sweeps away from the sun. How the tail appears to us depends on the relative positions of sun, Earth, and comet; occasionally a comet may even seem to have two tails pointing in opposite directions.

For some time it has been known that comets generally have two tails (Figure 4.4). Type I tails are clumpy and straight, whereas Type II tails are curved and more uniform in appearance. The tails are made of material that evaporates from the comet as it warms during its approach to the sun. Spectroscopic analysis of the light from the tails has identified their constituents. Type I tails are composed of gaseous ions such as CO^+ (carbon monoxide) and OH^+ (from the breakup of water vapor molecules); Type II tails, in contrast, contain many dust particles that were originally frozen into the icy comet but were then released as the binding ice evaporated. Under the pressure of solar radiation, the dust particles get blown away from the comet and can be seen by the sunlight they reflect. The ions are much smaller than the dust particles and hence far less affected by solar radiation.

Between 1951 and 1957, Ludwig Biermann of Göttingen studied Type I ion tails. In photographs taken over many days, Biermann followed the movement of dense clumps of ionized gas away from the head of the comet and deduced the changes in speed of these clumps. From these observations he inferred the magnitude of the force that had to be responsible and suggested that these effects could occur if the sun were continually emitting particles whose collisions with ions in the tails provided the driving force. Biermann was able to deduce the speed and the numbers of particles in this solar stream.

In 1958, a seminal paper was published by Eugene Parker of the University of Chicago. Drawing on Biermann's work, Parker set out a detailed mathematical model of the flow of solar particle radiation, associating it with other features of the sun such as its magnetic field, corona, and rotation. Parker's introduction of the concept of a solar wind was "received with a great deal of

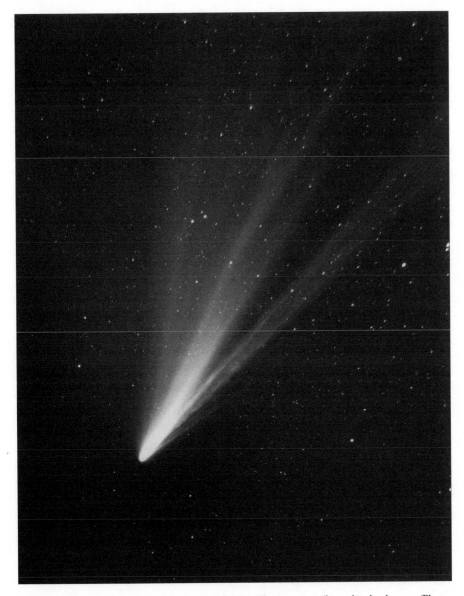

Figure 4.4. Comet West, photographed in March 1976. The separate tails can be clearly seen. (Photograph courtesy of Lick Observatory.)

skepticism, even disbelief" (as the NASA astronomer Jack Brandt described it), but the rest, as they say, is history. Parker's magneto-hydrodynamic model has been refined, and many features have been verified by satellites and long-range probes.

The first direct observation of solar wind particles came from the MIT detectors on *Explorer 10* in 1961, with many additional observations made during the following years. For example, the *Ulysses* spacecraft was launched in October 1990. It was aimed for a close swing by Jupiter in February 1992 so that its path was brought back around the sun during 1994–95, then out into the solar system, and it is due back again near the sun in 2000. At the farthest reaches of its orbit, it has obtained important measurements of the galactic cosmic rays (see Chapter 5), while solar wind and CR data are obtained much closer in. A special feature of *Ulysses'* orbit is that it is confined to being close to the ecliptic plane, but travels up to high latitudes.

Near the orbit of the Earth (at a distance of 1 AU from the sun), the solar wind moves with speeds of many hundreds of kilometers per second (450 km/sec is about 300 miles/sec) and contains about 5 protons per cubic centimeter—which means a flow of more than 100 million particles per square centimeter per second. The quantities vary considerably, and streams of particles sometimes move faster than 1,000 km/sec.

The solar wind is composed of particles with both positive and negative charge (protons and electrons), but because they occur in equal numbers, the stream is electrically neutral. The high temperatures (over 1 million degrees at the sun and half that by the time the solar wind reaches the Earth) ensure that the gas remains highly ionized. Within the stream is a magnetic field, so there is a magnetically complex region where the solar wind flows by the Earth and encounters the Earth's own magnetic field. The latter is distorted: compressed on the side toward the sun and drawn out on the opposite side. As a result, the Van Allen regions of trapped particles are also distorted (see Figure 3.7), as was discovered and mapped by spacecraft. Because of the sun's rotation, the solar wind does not simply stream out along straight radial directions like the spokes in a wagon wheel but follows curved spiraling paths, in what is often called the garden-hose effect. The Earth thus encounters the solar wind coming from a direction slightly ahead of the direct Earth-sun line. The solar wind has been detected by spacecraft as far out as 40 AU from the sun. It is thought that the solar wind terminates abruptly in a "shock," a complex boundary between the interplanetary and interstellar regions at greater distances, perhaps as far as 100 AU. The *Voyager 2* vehicle, launched in 1977, flew by Neptune in 1989. It is now more than 60 AU from the sun and heading out of the solar system. It is hoped that *Voyager* will still be working well as it traverses the boundary region.

What do we know of the processes that produce the solar wind? Ultimately, the energy comes from the sun's nuclear reactions, which sustain the high temperatures in the corona. Recent satellite observations of the sun, in the X-ray and ultraviolet regions, have shown coronal holes, regions of low density where the magnetic field of the sun is carried out with the escaping streams of particles. High-speed streams (with typical speeds around 700 km/sec) appear to be associated with coronal holes near the sun's polar regions; these regions appear dark on x-ray images of the sun. The sources of the slower streams (with speeds around 400 km/sec) are still not known precisely, but they seem to come mostly from the solar equatorial regions. The whole picture is strongly influenced by the sun's rotation and its intricate magnetic field. Thus the general qualitative features of solar-wind production seem to be understood, but the description is mathematically complex and much detail remains to be clarified.

The solar wind is a persistent though variable part of the quiet sun's output. Solar cosmic rays, however, represent one of several short-lived manifestations of the active sun (Figure 4.5). The solar wind and solar CR are composed of the same types of particles, electrically charged constituents of the outer solar layers. If you examined an individual particle, you could not tell whether it belonged in the solar wind or among the solar CR. Both groups are emitted sporadically, and it is difficult to correlate the variations with the solar cycle. What does distinguish these two groups of particles are their numbers and speeds: the cosmic rays have speeds and energies vastly greater than their solar-wind relations. A typical solar-wind proton will have a speed of 500 km/sec (about 0.2 percent of the speed of light) and an energy of 3,000 eV, while solar CR have energies measured in millions of electron volts and speeds above 15,000 km/sec (about 1/20 of the speed of light).

In the late 1950s, around the time of the IGY, several independent observations pointed to the clumpy emission of high-energy solar CR. Earlier, Forbush had noted an occasional decrease in the rates of his ground-level ionization chambers, but in February 1956 his station at Huancayo in Peru showed a huge increase in the CR intensity (as in Figure 4.5). At this location, near the magnetic equator, particle energies above 13 GeV were needed to get close to the Earth. At the same time, D. K. Bailey (of Page Communications Engineers and the National Bureau of Standards) noted major changes in the attenuation of radio signals. Absorption of the solar ultraviolet radiation maintains the degree of ionization in the ionosphere, 50 km or more from the Earth's surface. These ionized layers, in turn, control long-distance radio

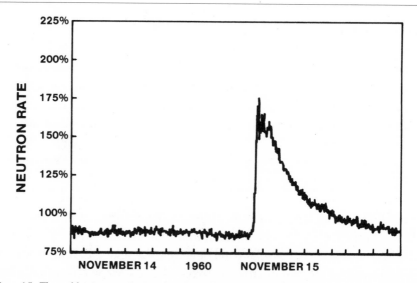

Figure 4.5. The sudden increase in cosmic ray activity associated with a large solar flare in November 1960, as shown by a sea-level neutron counter at Deep River, Canada. (Adapted from J. F. Steljes, H Carmichael and K. G. McCracken, Journal of Geophysical Research 66, [1961]: 1363; © American Geophysical Union.)

transmission. Changes in radio transmission thus provide a good way to monitor changes in the ionosphere, and Bailey's measurements pointed to the arrival of large numbers of ionizing particles.

Earlier high-altitude balloon flights had uncovered flare-related particles whose energies were too low to produce secondary effects detectable at sea level. The first clear observation of solar-flare protons was obtained by Kinsey Anderson, in Van Allen's department at the State University of Iowa, during a 1957 balloon flight at Fort Churchill near Hudson Bay. Other high-altitude CR observations together with the monitoring of radio transmission showed that radio blackouts occurred over large areas when many low-energy solar protons were channeled into the polar regions by the Earth's magnetic field. In the jargon of the field, the solar proton events produced polar cap absorption (PCA).

In a PCA event, the arrival of solar ultraviolet and X-ray photons increases the number of ionospheric electrons and interferes with radio communication, which, though a nuisance, does not damage radio receivers; transmission returns to normal as additional electrons reattach to atoms. We now have the ability to simulate something like a PCA, but the consequences are potentially serious. The detonation of a large hydrogen bomb above the atmosphere creates an intense fireball of gamma rays. From a typical altitude of 100 km, many of these photons enter the atmosphere and ionize numerous atoms. The sudden release of electrons triggers an intense electromagnetic pulse (EMP) that propagates through the atmosphere and can damage transistorized circuits much as an overload in a house can blow a fuse. The deliberate use of a nuclear bomb to incapacitate the communications network of an enemy is a cause of concern to the military. Shielding against EMP is not impossible but does carry a weight penalty that can be a serious problem for airplanes and satellites.

Large solar proton events directly related to major solar flares can be monitored optically. In a major flare, the light intensifies for about a minute and then rapidly dims. Similar flaring is generally detected at short wavelengths such as ultraviolet and X-rays. Other visible signs of solar activity can have an awesome scale. Figure 4.6 shows the giant solar prominence of June 1946. During an interval of only 98 minutes, the heated gas rose (at speeds of around 100 km/sec) to a distance comparable to the solar radius, or about 100 times the radius of the Earth. The role played by the sun's magnetic field in confining the motion of the charged particles in the glowing gas can be seen clearly in the sharp boundaries of the prominence and its looping shape.

The travel time from the sun to the Earth for electromagnetic radiation is 8 minutes, so the ionizing ultraviolet radiation and X-rays arrive as the flare is being seen and promptly increase the number of electrons in the ionosphere. Radio transmission is then noticeably affected. The charged particles travel

Figure 4.6. The solar prominence of 4 June 1946, photographed through a red filter to show hydrogen emission. The gas temperature varied between 25,000 and 100,000°K. Within an hour this prominence grew to become nearly as large as the sun itself. The shape of this prominence was strongly controlled by the local magnetic field. (Photograph courtesy of High Altitude Observatory.)

fast but still much slower than the light; their arrival at Earth follows after a delay that depends on their own speeds and on the solar-wind structure at the time. Typical delays are measured in hours, as Carrington and others had observed (but not understood). The geophysical effects on the ionosphere and radio transmission are too complex to pursue here, but the main features are now well established.

The occurrences of major flares are unpredictable, but their frequency follows the general sunspot cycle. High-altitude rockets have been used to detect the flare particles at times of expected increased solar activity. The advantage of rockets is the speed with which they can reach full altitude, far faster than balloons or satellites, which require protracted preparation and cannot be held ready for extended periods. On one occasion Carl Fichtel and Don Guss, of the NASA Goddard Space Flight Center, shot off rockets to observe the large solar flare of November 1960. Their rockets reached a peak altitude of 400,000 feet in a few minutes and were above 100,000 feet for about 5 minutes. Many solar protons were detected, as well as heavier particles, and thus these researchers were able to obtain a value for the hydrogen/helium ratio for solar CR. The presence of helium in the solar atmosphere had been known since its discovery by Joseph Lockyer in 1868, but estimating its abundance had been very difficult. The CR observations provided a reliable and new method.

Because CR particles are identified by their charges, the list of their abundances is usually called the charge spectrum. The charge spectrum of the solar cosmic rays, given in Table 4.1, is well established. The abundances, at least for

Table 4.1 Average solar cosmic ray abundances (relative to hydrogen = 100).

Nuclei	Abundance
Hydrogen	100
Helium	14
Lithium, beryllium, boron	<0.004
Carbon, nitrogen, oxygen	0.23
All heavier nuclei	0.037

energies above 10 MeV, vary by less than a factor of three between different events, and it appears that we are observing a representative sample of coronal material that has been accelerated.

Another characteristic of solar CR is their energy spectrum. Cosmic rays do not all have the same energy, and so do not all travel at the same speed. The number of particles with energy E (in MeV) depends on E and has been found to be proportional to E^{-a}. The minus sign indicates that there are fewer particles with larger values of E, and the value of a is usually in the range 2.5–3 for energies up to around 10 MeV, indicating a rapid decrease in the numbers of protons and other nuclei with increasing energy. Despite this general spectral shape, there is evidence for the presence of much higher energy particles in some flares.

When protons have energies above about 300 MeV, they can produce pions in their collisions with nuclei in the Earth's atmosphere. Some of the pions will be uncharged and promptly decay into high-energy gamma rays, such as were detected in the June 1982 solar flare.

The intensity or flux of cosmic rays is the number of particles that traverse a standard area (such as 1 square centimeter) each second. (Strictly speaking, intensity and flux are different measures, but this fine point can be ignored here.) During large flare events the solar CR intensity can exceed 100,000 particles/ cm²sec, with the total energy amounting to 10^{31} ergs. This is roughly equivalent to the energy released by 2.5×10^8 hydrogen bombs, but still only equal to the solar e-m radiation emitted in 1/300 of a second. (An erg is a very small unit of energy. It takes about a trillion ergs to boil a cup of water.)

Another component of the cosmic rays was discovered during 1972–73 by detectors aboard the *IMP-5, IMP-7,* and *Pioneer X* spacecraft. It was called Anomalous Cosmic Rays (ACR). These are particles that have not been ionized completely and whose energies are below the geomagnetic cutoff values; accordingly, they cannot be galactic CR, nor are they solar CR, which are both fully ionized and much more numerous.

By now, more than 25 years later and with the use of detectors on many spacecraft, including the *SAMPEX* polar orbiter and the *Ulysses* mission, ions of hydrogen, helium, carbon, nitrogen, oxygen, neon, silicon, sulfur, argon and iron have been detected, some singly ionized, others with net charges of +2 or +3, implying that one, two, or three electrons have been removed from initially neutral atoms. Some of the oxygen ions retain only one electron each. Energies range from several MeV/nucleon up through 100 MeV/nucleon, with the singly charged ions tending to have the lower energies.* The fluxes vary sufficiently through the solar cycle so that at times there are more helium than hydrogen ions. Other relative abundances, such as the C/O ratio, differ sharply from those of galactic and solar CR, and relative abundances are found to differ between the inner and outer heliospheres.

Though the complexity of the picture has become apparent, the main features of the theoretical description seem established. ACR particles are thought to have started as neutral atoms in the interstellar medium and then to have been partially ionized by the solar ultraviolet radiation, and perhaps also by the solar wind, which then accelerated them.

In contrast, the acceleration of solar CR particles takes place in the magnetic field associated with sunspots, through magneto-hydrodynamic processes. The mechanisms involved are probably similar to those that produce galactic cosmic rays, to which I turn in the following chapter. Although most of the experimental methods for detecting galactic CR are similar to those for detecting solar CR, the subject of galactic particles leads to a whole new range of astrophysical areas.

---

* Neutrons and protons, with closely similar masses, are often called nucleons. Nuclei are composed of nucleons. Heavier nuclei contain more nucleons. When comparing the energies and energy spectra of different types of nuclei, it is often more useful to compare energy/nucleon, which is a direct indicator of the speed of the nucleus.

5

COSMIC RAYS IN THE GALAXY

Cosmic ray studies extend far beyond the familiar region of the solar system to the galaxy, where vastly greater distances and different physical processes are encountered. The nature of the Milky Way was the subject of speculation and myth for centuries; then in 1609 Galileo's use of the newly invented telescope changed astronomy forever, freeing it from the limits set by the human eye. Galileo turned his glass in all directions and described his findings in his book *The Starry Messenger*, recording with surprise and exhilaration the nature of the Milky Way that was so suddenly and unexpectedly revealed (*Discoveries and Opinions of Galileo*, p. 49):

> I have observed the nature and material of the Milky Way. With the aid
> of the telescope this has been scrutinized so directly and with ocular
> certainty that all the disputes which have vexed philosophers through
> so many ages have been resolved ... The galaxy is, in fact, nothing but
> a congeries of innumerable stars grouped together in clusters. Upon
> whatever part of it the telescope is directed, a vast crowd of stars is im-
> mediately presented to view. Many of them are rather large and quite
> bright, while the number of smaller ones is quite beyond calculation.

From the lack of any perceived relative motion among the stars, astronomers long ago drew the correct inference that they were much farther away than the obviously wandering planets. Just how much more distant proved difficult to determine, and it was only as recently as 1836 that the first reliable distance measurements were made independently by Friedrich Wilhelm Bessel in Germany, Otto Struve in Russia, and Thomas Henderson at the Cape of Good Hope. Since then larger telescopes and more reliable measurements of stellar distances have led to the picture we have today, although it is still an incomplete one. Our sun was found to be just one among a system of a hundred billion stars (Figure 5.1).

To describe such great distances, new units are convenient. The popular and graphic unit is the light year, the distance that light travels in one year. Astronomers prefer the arcane parsec (pc), 3.1×10^{18} cm, and its larger multiples of one thousand (kiloparsec, kpc) and one million (megaparsec, Mpc). The parsec is based on triangulation with the Earth's orbit as a baseline; 1 parsec is equivalent to 3.26 light years.

Viewed from above (keep in mind that "above" and "below" are only relative terms out in space), the galaxy would appear roughly circular, with a diameter of 30 kpc (100,000 light years, or 10^{23} cm). The stars would appear to be generally concentrated in arms that trail out in a spiral pattern from the galactic center. The solar system is located about midway from the center, in one of the spiral arms. Viewed from the side, the galaxy would appear as a rather flattened disk with a thickness of around 1,000 pc (3×10^{21} cm), somewhat thicker with a bulge near the center and tapering slightly toward the outer edges. The concentration of stars in this disk appears as the Milky Way as we look from a position close to the galactic equator.

Between the spiral arms are traces of gas, mostly hydrogen and generally very cold. Within the spiral arms are more gas and occasional dense clouds that contain dust and a rich variety of complex molecules. The halo, a larger volume surrounding the disk, contains fewer stars and little gas. Because we are embedded in the disk, our view in some directions is obstructed by large amounts of interstellar dust, so we cannot, for example, use ordinary light to see to the galactic center. Knowledge of the galactic center comes from observations made at other wavelengths, such as infrared and radio.

This description of our galaxy has been pieced together from many sources. Some of them are direct observations of the Milky Way, but to a considerable

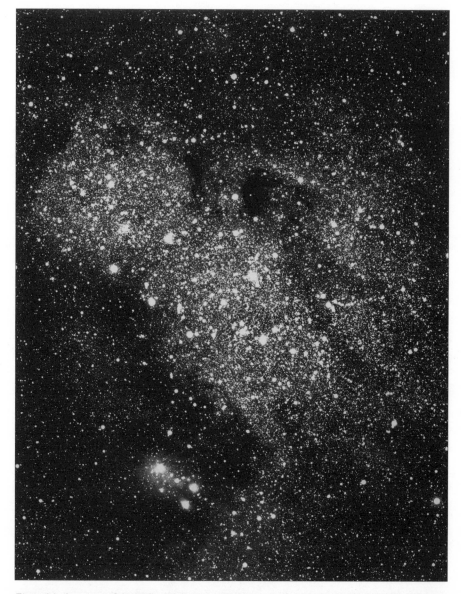

Figure 5.1. A section of the Milky Way galaxy, showing a dense cluster of stars in the constellation Sagittarius. (Photograph courtesy of Lick Observatory.)

degree we need to make comparisons with other galaxies seen through telescopes. The spiral structure of the Andromeda nebula, one of our closest galactic neighbors, is probably quite similar to our own. We can see other similarities, in the way in which our galaxy and Andromeda are rotating, in the distribution of the stars and dust, and in the way in which stars of different types are grouped.

Thinking about the structure of galaxies leads to a more general question: How do cosmic rays relate to the overall composition of the universe? Most cosmic rays are nuclei of atoms, and a central question in CR research has been the relative abundances of different particles. Simply compiling tables of abundances for solar and galactic cosmic rays, however, skirts the question. Are there any surprising features in such tabulations? Indeed, how would we know whether they were surprising unless we had a basis for comparison? To answer such questions, we need first to look more closely at the diversity of the atoms that make up the universe.

Each atom contains a positively charged nucleus surrounded by a cloud of negatively charged electrons; the result is an electrically neutral atom. It is customary to designate a nucleus by the number of units of its positive electric charge, which, in a normal atom, is equal to the number of surrounding electrons. This number, Z, is known as the atomic number, a useful index for compiling catalogs of elements and nuclei. Hydrogen, the lightest element, has a nucleus with $Z = 1$; helium has $Z = 2$; and the list continues through lead ($Z = 82$) and uranium ($Z = 92$). Beyond $Z = 92$ there are now over 25 types of nuclei (up to $Z = 118$) that have been produced by the bombardment of heavy nuclei targets with beams of particles from high-energy accelerators. These new transuranic nuclei are all radioactive with short half-lives and do not occur in nature.

An electric force attracts electrons to their parent nuclei; this force also holds atoms together in molecules, the result of sharing some of the outer electrons of each atom. Whether a conglomeration of molecules or atoms is a solid, a liquid, or a gas, depends on the properties of those atoms or molecules and on the temperature. At all temperatures, atoms and molecules are moving, even in solids. This internal motion becomes more vigorous as the temperature rises, until a solid melts to become a liquid and the liquid in turn evaporates to form a gas. Raising the temperature causes molecules in a gas to collide with increasing violence until they are torn apart into their constituent atoms. Further increases in temperature lead to more frequent and more violent collisions until eventually electrons get knocked off the atoms and the gas becomes a plasma, a cloud of positively charged ions, or nuclei, and negatively charged electrons.

The temperatures at which these changes occur differ from one type of atom to another, and also depend on the pressure. For example, at normal atmospheric pressure, H_2O molecules condense to form water at 100° C and solidify to form ice at 0° C. Hydrogen, usually thought of as a gas, is liquid at temperatures below −254° C. The surface of the sun has a temperature of about 6,000° C, and in that region about one hydrogen atom in a hundred thousand is ionized. Complete ionization of hydrogen requires a temperature of around 10,000° C, which is easily the case in the outer regions of the solar atmosphere (the corona), where the temperature exceeds 1 million degrees.

Because of this diversity of physical states, different methods of analysis must be used in compiling lists of abundances of different elements, depending on whether they are located on the Earth, on the sun, on other stars, or in the spaces between stars. Allowance must also somehow be made for the bulk of the universe beyond our immediate reach. Chemical analysis can be used for backyard dirt, the atmosphere, and the oceans, but what about the Earth's interior? Even the deepest mines go down only about 2 miles below the surface, a tiny fraction of the Earth's radius of 4,000 miles. By examining the sun's light in a spectroscope (Figure 5.2), we can identify different types of atoms from their characteristic spectral features, but what about the bulk of the sun that is far below the glowing photosphere?

There is no simple way to provide a definitive tabulation of elemental abundances. An obvious start is with materials that can be brought into laboratories, followed by those that can be examined only from a distance by the use of astronomical techniques. To proceed further one must resort to models, theoretical pictures that combine observed features with plausible assumptions. From such models one can deduce the overall composition of the Earth, the moon, the sun (Table 5.1), and more distant bodies; further observations may necessitate adjustment to the models.

Apart from materials found on Earth, three significant sources of material are available for analysis in laboratories: meteorites, dust particles, and samples returned from lunar missions. Countless objects, ranging in size from tons down to micrograms or smaller, orbit the sun, and their collisions with the Earth are matters of chance. Those that enter the Earth's atmosphere do so at high speed and heat up. The smallest objects burn up completely, but the largest (the meteorites) can survive to reach the ground, though they lose much of their outer material through vaporization as they plunge through the atmosphere. The composition of these objects, large and small, gives clues to the nature of the giant cloud of gas and dust from which the solar system is thought to have condensed about 4.5 billion years ago. Meteorites are classified as stony, iron, and stony-iron; one subclass, the carbonaceous chondrites, are regarded as the most primitive. Some micrometeorites, truly cosmic dust,

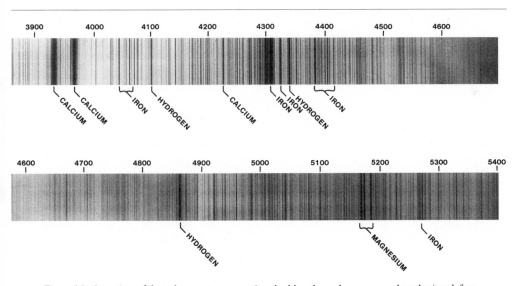

Figure 5.2. A section of the solar spectrum covering the blue through green wavelengths (top left to lower right). Some of the prominent lines produced by different types of atoms have been labeled. (Courtesy of the Observatories of the Carnegie Institution of Washington.)

have been detected from satellites and high-altitude research planes. Study of these tiny dust particles, typically no more than 1/40 mm in diameter, has given new insights into interplanetary composition and processes in the early solar system.

The third type of extraterrestrial material is lunar samples. During the six *Apollo* missions, a total of 382 kilograms (kg), or 838 pounds, of lunar material was returned to Earth, and the Russian *Luna 16* and 20 unmanned vehicles brought back 130 grams (g). Less than 10 percent of the *Apollo* material has thus far been distributed for analysis; the remainder is preserved at the NASA Johnson Space Center in Houston for future research.

In compiling a tabulation of elemental abundances, we must also include the material that is widely dispersed between the stars. That region is not totally empty but contains the interstellar medium (ISM), a thin gas composed of stray atoms and molecules with some dust particles. There are also some giant gas clouds in which the density is about a million times higher, with a rich variety of complex organic molecules. Still, the interstellar region is a remarkably good vacuum: a typical cubic centimeter contains only a single hydrogen atom. In contrast to these low gas densities, the best vacuum achieved under laboratory conditions is about 10^{14} atoms per cubic centimeter, at room tem-

Table 5.1 Solar composition.

Element		Atomic number (Z)	Number of atoms (relative to hydrogen)	Fraction of total mass
Hydrogen	H	1	1.00	0.708
Helium	He	2	9.8×10^{-2}	0.276
Carbon	C	6	3.6×10^{-2}	3×10^{-3}
Nitrogen	N	7	9.3×10^{-5}	9×10^{-4}
Oxygen	0	8	7.4×10^{-4}	8×10^{-3}
Neon	Ne	10	1.2×10^{-4}	1.7×10^{-3}
Iron	Fe	26	3.2×10^{-5}	1.2×10^{-4}

Source: Adapted from Roger J. Tayler, *The Sun as a Star* (Cambridge: Cambridge University Press, 1997), p. 24.

peratures. There are two ways of learning about the constituents of the ISM: some regions are dense and warm enough to emit measurable amounts of radiation, while others reveal themselves by their absorption of light from more distant stars whose spectra display characteristic features produced by the ISM.

From all these sources, the overall abundance of each element can be assessed. But an additional classification can be made that can lead to important insights into the operation of nuclear processes in the astrophysical setting. This classification is by isotopes within each element. Atoms were historically classified by their ability to combine (chemically) with other atoms. We now know that this property depends on the electrons surrounding each nucleus. Within each nucleus are positively charged protons and electrically neutral neutrons, held together not by any electrical force but by a much stronger nuclear force. Each chemical element is characterized by the number of protons in the nucleus, but many elements have nuclei with the same number of protons but different numbers of neutrons—the isotopes. The nucleus of the usual hydrogen atom is the proton, but a nucleus consisting of one proton and one neutron will still have only one orbiting electron. Chemically like hydrogen, this isotope is sometimes called heavy hydrogen but more often deuterium, and its nuclei are known as deuterons. The usual carbon atoms that we encounter are mostly carbon 12, ^{12}C, with six protons and six neutrons in each nucleus, but two other major isotopes also occur in measurable quantities, ^{13}C and a radioactive isotope, ^{14}C, whose interesting properties are examined in Chapter 13.

It should be evident by now that compiling a table of "universal" or "solar system" elemental abundances against which to compare cosmic rays involves a complex array of resources: the data from solar and stellar spectroscopy, meteoritic and geologic data, and a considerable knowledge of nuclear theory (some theory is needed because not all of the isotopic abundances have been measured). The modern theory of stellar evolution includes calculations of nuclear reactions that steadily convert hydrogen to heavier nuclei, producing all the atoms in the periodic table. The general features of the nucleosynthetic processes are well understood and provide the basis for filling gaps left in experimental tabulations. The results, summarized in Table 5.2, are based on the accepted standard compilations of Edward Anders (of the University of Chicago) and Nicholas Grevesse (of the Université de Liege in Belgium). Their complete list includes estimates of abundances of all isotopes in addition to the element-by-element totals.

Knowledge of elemental abundances among the cosmic rays, and within each element the proportions of individual isotopes, helps to identify the physical processes in which the particles were produced and accelerated, in stellar interiors or atmospheres or out in the vastness of interstellar space. During the particles' interstellar travel, some have suffered collisions and fragmentation that changed the relative abundances, and corrections must thus be made in extrapolating back from Earth-neighborhood observations to probable sources. The starting point, however, remains a set of well-determined elemental and isotopic abundances close to the Earth but above the atmosphere.

So far have the methods improved that, among the more abundant lighter nuclei, charges have been cleanly resolved up through iron ($Z = 26$) and nickel ($Z = 28$), as shown in Figure 5.3 for data from the *ISEE* spacecraft. Indeed, individual isotopes have also been resolved, as shown in Figure 5.4 for data from the *Ulysses* deep space probe. In the following chapters we will

Table 5.2 Solar system abundances.

Element	Relative number of atoms
Hydrogen	1.00
Helium	9.8×10^{-2}
Lithium, beryllium, boron	2.8×10^{-9}
Carbon, nitrogen, oxygen	1.3×10^{-3}
Neon through manganese	2.3×10^{-4}
Iron	3.2×10^{-5}
All heavier atoms	1.9×10^{-6}

Source: Adapted from tables given by Edward Anders and Nicholas Grevesse, *Geochimica et Cosmochimica Acta* 53 (1989): 197–214.

make use of this accumulation of observational results as we follow the attempts to reconstruct the history of the cosmic rays and identify their sources.

In describing galactic cosmic rays I will follow the pattern set with solar particles, beginning with their charge and energy spectra. With solar flare events, if the numbers of particles and their energies are known, one can calculate the total energy and see, for example, how small a flare is in comparison with the sun's regular radiant output. A survey of the corresponding properties of galactic cosmic rays will make it possible to construct a model for their production and their travel (propagation) in the galaxy. But new physical processes—quite negligible in the relatively minuscule solar system—become important because of the galaxy's scale.

Figure 5.3. Charge spectrum of heavy cosmic rays, showing the clean separation between charges. Charge resolution improves when only particles arriving within 15 degrees of the zenith are accepted. (Courtesy of Richard Leske; copyright the *Astrophysical Journal* 405 [1993]: 567.)

Figure 5.4. Isotope spectra for iron ($Z = 26$) and Nickel ($Z = 28$), as measured from the *Ulysses* space probe. In C, the scale has been enlarged to show the less abundant iron (Fe) isotopes 55, 57, and 58 more clearly. (Courtesy of James Connell, University of Chicago; *Astrophysical Journal* 407,[1997]: 61.)

A perpetual problem in observing cosmic rays is their paucity, and the situation gets worse when one is examining heavier particles or those that are the most energetic. Confidence in interpretations is based on the reliability of observations and thus on the number of particles detected, the certainty of their identification, and the accuracy of energy measurements. Detecting more particles is quite simply a matter of using larger apparatus for longer times, and thus the equipment used in CR studies has evolved from electroscopes a few inches in diameter to satellite-borne hardware weighing many tons and ground-based arrays of particle detectors covering some hundreds of acres.

Today satellites are the obvious route: get above the atmosphere and stay there for as long as necessary. Electronic detectors can relay information back to Earth, where analysis can take place in the comfort of one's office. In some cases, as in the manned *Skylab* and the space shuttle, photographic plates and some newer types of track detectors (made of plastics or glass) have been returned from orbit for processing and examination. But before satellites flew, and even long afterward, balloons have hoisted detectors high enough to map the major features of the cosmic radiation. The story of scientific ballooning deserves a book to itself: the development of special plastics to contain the lifting gas; the role of independent companies such as General Mills, Raven, Winzen, and Schjeldahl, which designed, made, and flew the balloons in the United States, and the far-sighted contribution of the U.S. Office of Naval Research, which funded the balloon programs for many years. The cosmic ray group at the University of Bristol was unique in that it designed, built, and flew its own balloons in the early days.

Ballooning continues, now coordinated by the National Scientific Balloon Facility in Palestine, Texas. Balloon loads have grown to tons, and volumes have increased to 40 million cubic feet. A 1-ton load can be lifted to an altitude of 130,000 feet. The science of balloon flying has become more sophisticated; early flights lasted only a few hours, but now flight can be extended to many days by taking advantage of the seasonal reversals in the stratospheric winds. Even more ambitious and also more successful have been the flights of the Japanese American Cooperative Emulsion Experiment (JACEE) carrying stacks of special photographic emulsions and directed toward the high-energy heavy particles, above 20 TeV (1 TeV = 10^{12} eV). Between 1979 and 1995, JACEE had two flights between Australia and South America, each lasting more than 5 days, and another five Antarctic circumpolar flights that each stayed aloft between 9 and 15 days.

All of us who have flown balloons will have our own stories about our flights (with good and bad endings). On many of our group's early flights we had a camera that took regular photographs of an altimeter and an Accutron watch to provide a record of the flight altitude. That watch survived a fall from 80,000 feet when the parachute failed and it was still running when the payload was found. One Bristol balloon came down across a rail line from London, and produced a stern letter from British Rail threatening to send a bill to cover the overtime pay for the removal of the plastic.

Satellites and space probes have also, of course, been vehicles for cosmic ray studies. When space on board is at a premium, the cosmic ray detector systems have to be small, and so only the most abundant types of particles will be recorded in significant numbers. Nevertheless, with probes reaching beyond the planetary orbits, valuable information is being obtained, with data returned by telemetry. Earth-orbital flights also offer exposure above the atmosphere, but with the possibility of returning track detectors for post-flight treatment and track analysis. Buford Price's group at the University of California at Berkeley had detectors on the Russian *MIR* space station during 1991–1995, some for over 2 years, others for more than 4 years. On the *Long Duration Exposure Facility (LDEF)*, packages from the Dublin Institute for Advanced Studies and the Space Science Department in Noordwijk, were in orbit for nearly six years.

Hoisting a scientific load, whether by balloon or satellite, requires more than simply making sure the hardware functions as planned in a hostile environment. The particles detected and the ease of their identification depend on the location of the balloon flight or the orbit of the satellite. To understand the importance of these factors, it is necessary to probe a bit deeper into the method of identifying particles.

In Chapter 2, I described several methods of identifying particles, all based on measurements of the energy lost by a particle traversing the detector. The rate at which the particle loses energy depends on the particle's charge and speed proportionally as Z^2/v^2. As the velocity (v) increases, the energy loss rate drops. There is a minimum loss rate, however, since the speed of a particle cannot exceed the speed of light in a vacuum. (This is a basic postulate of Einstein's special theory of relativity, well confirmed by experiment.) As energy increases, the energy loss rate goes through a minimum value for speeds around 0.85 times the speed of light, and then slowly increases again.

The precise behavior of the energy loss rate at higher energies depends on the nature of the detecting material. A typical set of response curves is shown in Figure 5.5. To compare particles with different masses but the same speeds, we usually refer to kinetic energy per nucleon, MeV/n, where nucleon refers to the number of particles in the nucleus, regardless of whether they are protons or neutrons. A single proton and an alpha particle (containing two protons and two neutrons) that have the same speeds will have the same energy per nucleon, although the actual (total) kinetic energy of the alpha particle will be four times that of the proton.

Identifying a particle generally requires two independent measurements that depend in different ways on the particle's charge *(Z)* and speed. For particles with energies above about 1.5 GeV, the ionization loss rate (as seen, for example, in a scintillation counter) depends almost entirely on Z^2 (Figure 5.6). Provided the energy is in this range, a single measurement of ionization serves to identify the particle by its charge. This is where the Earth's magnetic field comes into the picture. When a detector is lifted by a balloon over Texas, only particles with energies greater than about 1.5 GeV will be let in by the magnetic field, so one can concentrate on measuring ionization. For a balloon flight originating in Guam, closer to the magnetic equator, the magnetic cutoff energy is much higher. On a flight from the northern United States or Fort Churchill in Canada, on the other hand, the cutoff energies are low. For particles with energies below about 1.5 GeV, identification can be made from several successive measurements of energy loss, sometimes in conjunction with deflection in a calibrated magnetic field or with a speed-sensitive Čerenkov detector. During a satellite flight the cutoff energy varies with each orbit, depending on the spacecraft's position. The arrival time must be noted for each

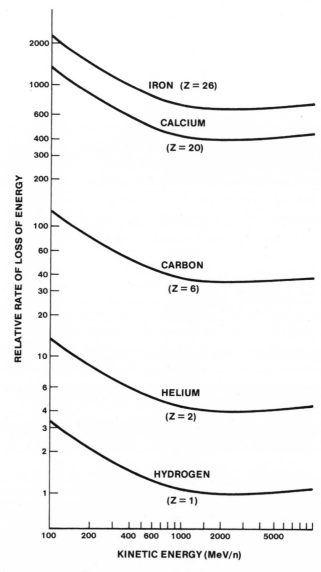

Figure 5.5. Rates of energy loss for several particles with different charge, Z. The curves show the Z^2 dependence in their vertical spacing. From 100 to around 800 MeV/n, the energy loss rate decreases with the square of the particle's speed; after a broad minimum, the loss rate increases only slowly.

particle so that corrections can be made later to allow for different times spent by the satellite in regions of differing cutoff energy.

In Chapter 3, I described how in the 1930s scientists came to understand the role of the Earth's magnetic field when the latitude and east-west effects were discovered. These two observations established the essential nature of most cosmic rays as positively charged particles and ruled out Millikan's tenaciously held view that the primary CR were gamma rays. Both direct and indirect evidence pointed to protons and helium nuclei as the primary particles.

The discovery of heavier particles came in 1948, as university research resumed after the end of World War II. Photographic emulsions and cloud chambers were carried by balloons to altitudes above 90,000 feet, where they recorded particle tracks that were clearly much more heavily charged than protons. In papers published between 1948 and 1950, groups at the Universities of Rochester and Minnesota reported their discoveries. At the northern latitudes of the balloon flights, low-energy CR were seen, some so slow they came to rest in the emulsions. The Ilford and Eastman Kodak emulsions were not yet sensitive enough to record the fastest alpha particles, but many heavier particles could be identified, showing charges of $Z = 6$ or greater. The presence of alpha particles was established initially from tracks in a cloud chamber.

With these discoveries, the emphasis in CR research shifted to the charge spectrum—the determination of the abundance of each type of particle (Figure 5.7). The comparison of this spectrum with other astrophysical abundance tabulations remains a central focus of CR research. A brief review of some aspects of particle behavior will be useful as background to the interpretation of observations.

When two protons collide, they cannot be fragmented. Mesons can be created from the kinetic energy of the collision, and charges may be exchanged so that neutrons emerge instead of protons (with mesons taking care of the electric charge balance). The situation is more complex when heavy nuclei collide, for either the projectile or the target or both may be fragmented. This process can be described by three quantities. The mean free path (mfp) is the average distance a particle travels before it has a collision. Clearly, the mfp is

Figure 5.6. Left and right: Tracks of fast primary cosmic rays showing the dependence of their ionization on their charge. The number of grains and slow secondary electrons (δ-rays) increases with Z^2. (Photograph copyright Powell Fowler Perkins/Science Photo Library.)

Ne.	Na.	Mg.	Si.	Ca.	Ti.	Fe.
Z=10.	Z=11.	Z=12.	Z=14.	Z=20.	Z=22.	Z=26.

⊢—50μ—⊣

Figure 5.7. Charge spectrum of the cosmic rays, from $Z = 1$ through 28, as observed at the Earth (solid line), compared to their abundance in the solar system (dashed line).

large in interstellar space where target atoms are far part, but it is much shorter in the atmosphere and becomes progressively shorter at lower altitudes as the air density increases.

The second quantity is the cross section, the effective target size (area), which depends on how many nucleons are in the nucleus. The volume of a nucleus is roughly proportional to the number (A) of nucleons (protons plus neutrons) it contains, and therefore the radius must be roughly proportional to the one-third power of A. The cross-sectional area will be proportional to the square of the radius and thus to the two-thirds power of A. The mfp and cross section are closely related and determine the frequency with which particles collide. A third quantity is needed to describe the types and numbers of fragments produced in each kind of collision.

Collisions take place both in interstellar space and in the atmosphere. In the early balloon flights, even those as high as 90,000 feet, the overlying atmosphere contained only about 1 percent of the total atmospheric atoms. In such a small amount of overlying atmosphere only about 10 percent of the CR protons will have collisions, but this is not the case for the heavier particles such as nuclei of carbon (C), nitrogen (N), oxygen (O), and iron (Fe), with their large cross sections. Many of these heavier particles will have collisions, and their fragments will include protons, alpha particles, and the nuclei of the light atoms, lithium (Li), beryllium (Be), and boron (B). When cosmic ray Li, Be, and B (the so-called L nuclei) were discovered in the 1948 emulsions, the immediate question was: Were these true primaries that had survived through interstellar space and then through the atmosphere down to balloon altitude, or were they secondaries, fragments from the collisions of C, N, and O (the medium, or M nuclei) or even heavier nuclei? The importance of this question lies in the general scarcity of the L nuclei in nature, both on Earth and more broadly in the galaxy.

The abundance of the L nuclei relative to the next heavier group, the medium, or M, nuclei (C, N, and O), became the focus of dispute in the 1950s, as Bernard Peters (in Rochester, New York) maintained that the observed L nuclei were all atmospheric secondaries while Peter Fowler and his colleagues (in Bristol, England) claimed a significant presence of primary L nuclei, even after allowing for production in the atmosphere. The resolution of this controversy came from a 1958 balloon flight carrying emulsions to a new record altitude of 120,000 feet. The residual overlying atmosphere during this flight was less than one quarter of 1 percent of the atmosphere, and the corrections for fragmentation were small enough that the results were beyond dispute. The flight location had been chosen in such a way that only relativistic particles would be detected, to reduce the uncertainties in particle identification. The group at the U.S. Naval Research Laboratory under Maurice Shapiro (who had been a student of Compton) reported their results in 1961 at the Kyoto CR conference: there were clearly primary L nuclei, although not as many as had sometimes been suggested.

With the presence of primary L nuclei at the top of the atmosphere firmly established, there was steady improvement over the next few years in methods of particle identification, especially through the use of a combination of Čerenkov and scintillation detectors. Introduced in 1956 by Frank McDonald at the State University of Iowa, this combination has been widely used and has gradually clarified the charge spectrum through iron, the twenty-sixth element of the periodic table. Beyond this point, the situation was unclear for many years. On Earth the periodic table of the elements extends up to uranium ($Z = 92$), but the abundances drop steeply for atoms heavier than iron ($Z = 26$). If nuclei heavier than iron were among the cosmic rays, it was therefore anticipated that they would be rare. Some CR experiments had shown particles that seemed to be heavier than iron, but problems with their identification left researchers skeptical and the picture unsettled.

In the mid-1960s two quite different sets of observations demonstrated the existence of heavier particles among the cosmic rays. The first of these introduced a new technique to CR particle studies. Robert Fleischer, Buford Price, and Robert Walker at the General Electric Research and Development Center had been studying particle tracks formed in certain types of crystals and later revealed by chemical etching. Some clear plastics such as cellulose nitrate and Lexan were also found to register tracks of heavily ionizing particles. The GE researchers showed that measuring track dimensions could lead to particle identification, much as with tracks in photographic emulsions. The group's surprising discovery of heavy CR came from crystals extracted from meteorites, in which Fleischer, Price, and Walker found particle tracks that could only have been produced by particles much heavier than iron. Tracks such as these cannot be the products of radioactive inclusions in the meteorites, and the conclusion was inescapable—that the tracks had been produced by cosmic rays and had slowly accumulated throughout the meteorites' lifetimes, estimated to be in the hundreds of millions of years. The number of tracks found was in good agreement with what would be expected if the relative abundances of these heavy particles followed the generally known elemental abundances and if the CR intensity through those hundreds of millions of years had been about the same as today. The meteorite observations thus significantly extended the CR charge spectrum and also suggested that CR flow in the galaxy was not simply a recent phenomenon.

While the meteorites provided a record of heavy cosmic rays covering very long times, the discovery of heavy particles among today's cosmic rays came from photographic emulsions carried on a balloon flight in 1967. Peter Fowler of Bristol University increased the total collecting area of his emulsion package to more than 4 square meters and found many tracks definitely attributable to trans-iron nuclei, with one track possibly produced by a particle with a charge as large as 90, perhaps even uranium.

Through the next five years, balloons carried payloads of increasing size— flying barndoors, as they were called. The large areas that had to be scanned for tracks required a collaborative effort (an increasing trend in physics), initially involving Fowler, the GE researchers, and our group at Washington University in St. Louis. Complementing the emulsions were the new plastic detectors such as those made of Lexan; these had the advantage of not responding to lightly ionizing particles, which in a photographic emulsion would almost swamp the rare tracks of the heavy particles. A flight payload usually contained a sandwich of several layers of emulsion interleaved with plastics of different sensitivities. During a typical balloon flight 500 tracks of heavy particles would be found, about one in every 200 cm². Locating these tracks was a typical needle-in-the-haystack problem.

Once again, a technical advance yielded further significant improvements in the delineation of the charge spectrum for heavily charged particles. From an experiment conceived in 1969, it took 10 years before a consortium of groups at Cal Tech, the University of Minnesota, and Washington University saw its 6-square meters electronic detector launched on the third *High Energy Astronomical Observatory (HEAO-3)* satellite (see Figure 2.7). A similar system was sent up by Fowler in the British *Ariel VI* satellite. These systems were designed to examine the charge spectrum upward of about $Z = 20$; the *HEAO* satellite also carried detectors for a French-Danish collaboration to cover the lighter particles, those with charges below 26. *HEAO* had a larger collecting area and longer exposure time than the older emulsions and plastics, and also provided much better charge resolution.

Among the very heaviest particles, up through platinum ($Z = 78$) and lead ($Z = 82$), the charges of individual nuclei have now been measured with a resolution of half a charge unit. Price, one of the pioneers in the etch-track method and now at the University of California at Berkeley, had packages of etch-track detectors carried on the Russian space station *MIR* during 1991–1995. Figure 5.8 shows the spectrum of the heaviest particles observed by Price's group. The Dublin-Noordwijk group, with their detectors for several years on the NASA *Long Duration Exposure Facility (LDEF),* did not have as good charge resolution, but their results show a similar grouping.

The best charge spectrum today comes from a combination of data from balloon and satellite experiments. Hydrogen and helium nuclei are so plentiful that adequate statistics have been accumulated during balloon flights, which have also produced excellent spectra up through iron. For the less abundant species and especially for the very heaviest particles, the extended duration of satellite and space probe exposures is needed.

To summarize, what we find is a charge spectrum that shows a persistent trend of diminishing abundances with increasing mass (charge), but with some important features superimposed. From carbon ($Z = 6$) through nickel ($Z = 28$), particles of even charge are more plentiful than adjacent particles with an odd Z (see Figure 5.7). This odd-even effect is well understood on the basis of nuclear properties, and it persists through the heaviest particles detected, where the abundances are dominated by the more plentiful even-charged particles, with a peak around $Z = 78$ (platinum), a smaller peak at $Z = 82$ (lead), and traces around $Z = 92$ (uranium). The rarity of the heaviest particles is clear: for every 3 billion protons, there is only one particle with Z greater than 65, and those with Z above 74 are 30 times less frequent still. The relative proportions of different particles within the primary CR beam can be seen in the measured fluxes listed in Table 5.3 for particles with energies above 1.5 GeV/nucleon.

When we want to go beyond the elemental abundances and examine the proportions of individual isotopes, we can still make use of ionization mea-

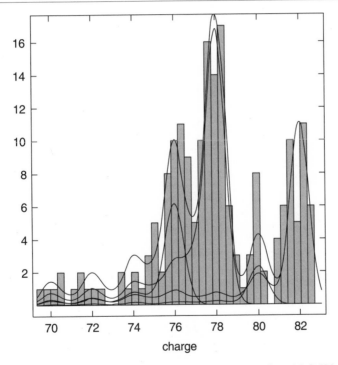

Figure 5.8. Charge spectrum for the heaviest CR particles, with energies above 0.9 GeV/nucleon. The scale on the vertical axis shows the actual number of particles, observed in detectors on *MIR*, 1991–1995. Six additional particles heavier than lead ($Z = 82$) were also observed. The curves represent different theoretical fits. (Courtesy of Andrew Westphal, University of California, Berkeley; reprinted by permission from *Nature* 396 [1998]: 50; © 1998 Macmillan Magazines, Ltd.)

surements. Ionization energy loss rate (MeV loss per cm) depends on the particle's charge and speed but not on its mass. A particle can travel until its kinetic energy is exhausted; the distance this takes is called the range, and this quantity does depend on the particle's mass. From theory, ranges can be computed in terms of initial energy for particles of different mass. Because they have different masses, different isotopes will have different ranges for the same initial energy. The relation between range and energy is the basis for the best measurements of particle masses and thus for discriminating between isotopes within a given element. Because important information is conveyed by the relative abundances of different isotopes, considerable technical ingenuity has been invested in designing experiments in which particle ranges are measured after initial measurements of ionization (energy loss rate). The range of a fast particle increases rapidly with its energy, so it becomes impracticable simply to build larger and larger detectors to stop higher-energy particles; after a while the required detector size would exceed the mean free path and too many particles would suffer collisions and fragmentation well before they could be brought to rest. The range method therefore works best with the less energetic particles, typically those with kinetic energy below about 400 MeV/nucleon.

Table 5.3 Fluxes of cosmic ray particles.

Type of nucleus	Flux (with energies over 1.5 GeV/nucleon) (particles/m^2·sec)
Hydrogen	640
Helium	94
Lithium, beryllium, boron	1.5
Carbon, nitrogen, oxygen	6
Nuclei with $Z = 9-25$	1.0
Iron ($Z = 26$)	0.24
Cobalt, nickel ($Z = 27, 28$)	0.01
All nuclei with $Z > 28$	0.003
All nuclei with $Z > 65$	2×10^{-7}

Several recent measurements of isotope abundances in cosmic rays have been obtained by groups at the University of California at Berkeley, the University of Chicago, and the University of New Hampshire. Among the beryllium (Be) nuclei, for example, 55 percent were found to be ^{7}Be and 39 percent ^{9}Be; radioactive ^{10}Be made up the remaining 6 percent. The precise proportion of ^{10}Be turns out to be of great importance in the theory of CR propagation. Carbon was found to be overwhelmingly ^{12}C; nitrogen had roughly equal amounts of ^{14}N and ^{15}N. Among the heavier nuclei, the best recent isotopic data come from the Cosmic Ray Isotope Spectrometer (CRIS) aboard the *Advanced Cosmic Explorer (ACE)* that was launched in August 1997. Isotopes of iron, cobalt and nickel have been cleanly separated.

Some of these isotopes were injected at the original sources while others have been produced by collisions en route to the Earth, and the proportions carry interesting information regarding sources, acceleration, and propagation history. We shall return to these data in Chapter 9 when discussing theories of the origin of cosmic rays.

After the nuclei and their isotopes, the next category in the catalog of CR particles includes electrons, once thought to constitute the majority of the cosmic rays, but now known to amount to less than 2 percent of the number of protons. At first this might seem surprising, for if CR protons had started out from atomic hydrogen, there should be equal numbers of protons and electrons; thus their different abundances must be explained in the model for CR production and propagation.

Definitive detection of primary electrons was achieved in 1961 by James Earl of the University of Minnesota, using a cloud chamber, and by Peter Meyer and Rochus Vogt of the University of Chicago, using electronic detectors. High-altitude flights are required to avoid the many secondary electrons that emerge from interactions in the atmosphere. I will discuss the electrons in more detail in Chapter 10, but there is one aspect that deserves mention here. Since all electrons have identical masses, interest is confined to their energy spectrum and to the relative proportions of electrons with positive and negative electric charge—which introduces yet another new topic.

Atoms with their positively charged nuclei and negatively charged electrons constitute the matter that makes up our world. In principle, we could imagine another world in which nuclei of negative charge would be surrounded by positrons—a world of antimatter, made up of antiparticles. Antiatoms would have the same properties as familiar atoms; they would combine to form antimolecules and could even be the basis for antilife, physically no different from our own. From a distance, there is no way to distinguish between stars and antistars.

The theoretical foundation for antimatter was set out by P. A. M. Dirac in his quantum theory calculations in 1929. He suggested that particles and antiparticles have identical masses but opposite charges. A property of such mirror-image sets of particles is that they can be created in pairs (of particle and antiparticle) from energy (Figure 5.9). When they meet they annihilate, their electric charges neutralizing each other and their masses being converted back to energy. In both creation and annihilation, energy and mass convert into each other, via $E = mc^2$, as first suggested by Einstein. In electron-positron annihilation, the energy emerges in two or three X-ray photons, and in proton-antiproton annihilation the result is gamma rays or mesons.

This scenario is not simply a science-fiction product of the imagination. The predictions of Dirac's theory have been repeatedly confirmed. In 1932 antielectrons (positrons) were discovered among the secondary cosmic rays, and antiprotons were first seen in 1955 in an experiment at the new Bevatron accelerator at the University of California's Radiation Laboratory in Berkeley. Not unexpectedly, antiparticles enter into CR physics. Positrons are produced when sufficiently energetic photons (above 1 MeV) convert into electron-positron pairs; they also come from the decay of mesons produced in nuclear collisions. Much more energy is needed to produce a proton-antiproton pair, and as a result antiprotons are rare. The detection of antiparticles among the cosmic rays is thus a clue to physical processes that have occurred at their origin or during their travel. We do expect some antiprotons to be produced by CR collisions with interstellar atoms. However, distinguishing between particle and antiparticle cosmic rays requires techniques that differ from those previously described, because particle ionization depends on Z^2 and we find the same energy loss rate whether the particle has a positive or a negative charge.

Figure 5.9. Electron pairs produced by gamma ray photons. At the point where a gamma ray is converted into an electron-positron pair, the two particles are very close together. Their separate tracks can be seen after a short distance. (Photograph courtesy of Peter Fowler, University of Bristol.)

There are two ways to identify antiparticles. One method utilizes the bending property of a magnetic field that forces charged particles to follow curved paths. The degree of curvature depends directly on the field strength and the particle's charge, and inversely on the particle's mass and speed (its momentum). Because the charges can be positive or negative, the direction of curvature immediately shows the sign of the charge. This was the method used by Anderson and Neddermeyer when their they discovered the positron during the attempts to measure particle momentum by observing track curvature in a cloud chamber (see Figure 3.4).

The other method of identifying antiparticles rests on their property of annihilation in collisions with their particle counterparts. If the total energy of particles coming from a collision is much more than the incoming kinetic energy, then it is reasonable to assume that the added energy has come from annihilation. Emilio Segre and his colleagues at the University of California at Berkeley, used the beam of 5.6 GeV protons to produce antiprotons in collisions. They did not detect photons but rather the mesons created from the considerable annihilation energy. The annihilation of antiprotons releases nearly 2,000 times more energy than does electron-positron annihilation and provides an unmistakable signature.

In practice, the search for primary cosmic antiparticles requires the magnetic curvature method, but magnets and their power supplies weigh a lot and there have been only a few sets of measurements. The first reported observation of CR antiprotons came in 1979 from a balloon experiment carried out by a group at NASA's Johnson Space Center in Houston. That group's system did include a superconducting magnet for measurement of trajectory curvature, along with a Čerenkov detector and a shower counter, and was aimed at particles with energies of around 10 GeV. The flux of antiprotons they reported was around 5×10^{-4} times the corresponding proton flux, rather higher than the value of $(1-2) \times 10^{-4}$ expected to result from regular CR collisions with interstellar gas during transit to Earth. The rejection of background events that mimic antiprotons is critical in such research. Two more recent systems have had improved detector systems to hunt for antiprotons, the Isotope Matter Antimatter Experiment (IMAX) and the Balloon-borne Solenoidal Spectrometer (BESS). At energies around 400 MeV to 1 GeV, there are no more than 1 antiproton for every 10,000 (regular) protons, in excellent agreement with the calculated value for secondary antiprotons from in-transit collisions. The important implication of these observations is that no source of extragalactic antimatter is required nor is there any need for bizarre sources within our own galaxy.

For antielectrons—the positrons—the situation is similar. The High Energy Antimatter Telescope (HEAT) of Greg Tarle (University of Michigan) and Simon Swordy (University of Chicago) weighs 5,000 pounds with its magnet and was carried by a balloon to a height of 120,000 feet during two flights in 1994 and 1995. They found about as many low-energy positrons as expected, but rather more at higher energies. They noted that the excess was not particularly large and could result from experimental errors, which can sometimes be very hard to track down. These are important measurements to make, and the exclusion of background and instrumental effects is difficult.

Positrons and antiprotons have been created under laboratory conditions, and some have also been seen among the cosmic rays. Might there not also be heavier cosmic antiparticles, considering the violence of cosmic eruptions with their vast quantities of energy? There have been searches, but no such antiparticle has yet been seen. The search continues, with the Alpha Magnetic Spectrometer (AMS) designed to be able to detect heavy antimatter nuclei. If all goes well, it will be on the International Space Station in 2002. But before leaving the subject of antiparticles, we can take note of an astronomical application and a practical use.

When an electron and a positron are about to annihilate they may first form positronium, which is very much like a hydrogen atom but with the positron playing the role of the usual proton as the nucleus; Martin Deutsch of MIT was the first to observe positronium, in 1951. Annihilation follows rapidly and produces two or three characteristic X-ray photons that are easily detected. If only two photons emerge from the annihilation, they each have an energy of 0.511 MeV, and these have been observed by X-ray astronomers. The process of annihilation may sound like an unearthly phenomenon, but in fact it is now a diagnostic tool of radiologists, who inject into a patient radioactive tracer isotopes that emit positrons. The tracers spread through the body, and the annihilation photons from the emitted positrons are then detected and mapped to produce images resembling X-ray photographs. This technique is called positron emission tomography (PET).

In this chapter and the last I have cataloged the types of CR particles. The measurements that identify these particles have also been used to measure their energies, and I turn next to a discussion of the energy spectrum of the CR population.

6

THE ENERGY SPECTRUM

As we seek to understand the overall astronomical picture of the cosmic rays, we need to know how much energy the cosmic rays carry individually and together. Cosmic ray particles do not all have the same energy. Their spread of energies is termed the energy spectrum, and the shape of this spectrum can tell us a lot about the physical processes through which the particles have been accelerated to such great energies and also about their subsequent travels before we detect them close to the Earth. If we measure the energy carried by individual particles and know the numbers of particles with different energies, we can calculate the total energy in the cosmic ray beam.

The energy picture of the CR population is not static. All cosmic rays lose energy (generally only a little) as they travel through interstellar space, but some particles can have catastrophically destructive collisions with interstellar nuclei, fragmenting both target and projectile and creating mesons. Some cosmic rays leak out of the galaxy. If the overall population and energy content of cosmic rays are to remain at a steady level, then new particles must be injected to replace lost ones. Acceleration processes must maintain the energy spectrum. Once the energy gains and losses of cosmic rays are known, the energy budget can be determined, and only then can we hope to see which among the possible CR sources possess the necessary reservoirs of energy to inject new particles.

It will be helpful in this discussion to clarify the sometimes confusing sets of units used for denoting particle kinetic energies—a situation that has been made more complex by the fact that CR astrophysics brings together subfields that developed separately, each with its own jargon. Through the preceding chapters we have become familiar with the use of multiples of the electron volt, such as MeV (10^6) and GeV (10^9). For particles with much more energy, the multiples TeV (10^{12}) and EeV (10^{18}) can be found in the professional cosmic ray literature. However, for simplicity I will often specify them in eV with powers of 10, such as 3×10^{18} eV. Knowing a particle's kinetic energy, we can calculate its velocity, a quantity needed for some calculations.

Where the cosmic ray particle is a nucleus larger than a single proton, the situation is somewhat more complicated. For example, a regular carbon nucleus will contain 6 protons and 6 neutrons. A kinetic energy of 600 MeV, for example, is thus shared by 12 nucleons, and the kinetic energy is expressed as 50 MeV/nucleon, and the velocity of the nucleus is the same as that of a single 50 MeV proton.

Because the masses of the cosmic ray particles are so small—one proton has a mass of about 10^{-23} gm—the electron volt and its multiples are fine for describing their kinetic energies. A larger scientific unit of energy, one that predates the electron volt, is the erg; the relation between erg and electron volt is expressed as 1 eV $= 1.6 \times 10^{-12}$ erg. Most of us, however, are more familiar with the calorie, as a measure of the energy content of various foods or of the energy "burned off" during exercise; 40 billion ergs is the equivalent of 1 diet calorie.

Different methods must be used for measuring the energies of particles in different energy ranges. One widely used method measures the energy that a particle loses by ionization as it goes through a detector. In this process, energy is transferred to the electrons in the medium being traversed and some of those electrons receive sufficient energy to be removed from their parent atoms. The faster a CR particle moves, the less time it spends near any atom with its electrons and so the less time it has to transfer energy. Ionization energy loss thus decreases for faster particles, but it reaches a minimum value when the CR particle has a speed about 85 percent of the speed of light. For higher speeds, there is a slow increase in ionization, which is well explained in the full theory of energy loss.

For particles with energies under about 300 MeV, the distances they can penetrate in solid matter before being brought to rest (their ranges) are short enough that they can be measured in conjunction with ionization. Particles with intermediate energies, between about 300 MeV and 1,500 MeV (1.5 GeV), will usually penetrate completely through detectors, but their energy loss in transit can be measured. In some transparent materials the cosmic ray makes the atoms scintillate and the resulting light can be measured. In other transparent materials and especially gases, the cosmic ray particles can travel faster than the speed of light in that medium and emit Čerenkov radiation. (This does not violate the basic tenet of relativity—it is only the speed of light in a vacuum that cannot be exceeded.) Scintillation and Čerenkov counters are widely employed to measure particles with these intermediate energies.

For particles with energies between 1.5 and 100 GeV, there are two current methods of measurement. One is based on ionization. Slow and fast particles of different charge (Z) and speed (v) can suffer the same energy loss (recall the dependence on Z^2/v^2), but the ambiguity is removed by using a Čerenkov detector, which identifies the fast particles through the difference in response on Z and v. Use of the relativistic rise in energy loss is generally confined to gaseous ionization detectors, since the increase in solid scintillators is too small to be useful. The other method in this energy region is the use of a magnetic spectrograph, which records the curvature of a particle's path in a strong magnetic field. The more energetic the particle, the less curved will be its path.

Up to around 100 GeV, all the methods of measurement are based on the signals generated along a particle's path. Above this energy, ionization and track curvature do not exhibit measurable changes with increasing energy; for example, scintillation detectors cannot distinguish between 500 and 5,000 GeV, and the magnetic fields available are not strong enough to produce measurable curvatures for these particles. Consequently, less direct methods, based on nuclear collisions, are called for. This at once introduces a problem: we can measure the ionization signal of almost every particle that passes through a detector, but not all particles have collisions that can be analyzed. Many particles will not be so obliging as to have their collisions in our detectors or in some other observable way; in the end, statistical corrections must compensate for the unobserved particles.

Which features of CR collisions can be exploited to assess the energy of the incoming particle? As collisions become more violent, the results gradually change in character, as seen in the "stars" in Figure 6.1. When an incident particle has only a few hundred MeV, some of this energy is transferred to the target nucleus, where it can promptly knock out a few particles. Some of the energy rattles around and is distributed, quite unevenly, among the remaining nucleons. Occasionally one of these nucleons accumulates enough energy to escape from the nucleus (it is said to evaporate), and, for example, the track of an emerging low-energy proton may appear. Perhaps a group of four nucleons may come out together as a heavily ionizing alpha particle, revealed by its dark track. We may even see short and thicker tracks of heavier fragments from the target. The characteristic time scale for evaporation is incredibly short by everyday standards, around 10^{-20} sec.

When we examine the shape of the energy spectrum, we find, as with the solar CR, that the number of particles decreases with increasing energy, proportionally as E^{-a} (Figure 6.2). The flux of CR particles decreases as their energies increase, so that detection of higher-energy particles requires larger and larger collecting areas. Very large-area detecting systems (each of several square meters) have been used on stratospheric balloon flights: the JACEE system detected 656 protons with energies above 6 TeV and 414 helium nuclei above 8 Tev/nucleon during 15 balloon flights, and a Russian-Japanese team had four flights.

As the energy of an incident CR particle increases, it can generate mesons in a collision—mostly pions but also a few even more exotic particles—see Figure 6.1. In emulsions, we see the tracks of shower particles, thin tracks with

Figure 6.1. Nuclear disintegrations produced by protons of increasing energy: Left: 570 MeV. Eight nuclear fragments each leave a heavily ionizing track, but no mesons were produced in this collision. Middle: 30 GeV. A heavy nucleus in the emulsion (silver or bromine) is torn apart by the incoming proton. Twenty-two heavily ionizing fragments emerge along with nine newly created "shower" particles, all pions. Right: 200 GeV and 300 GeV. So much energy is available that "jets" of mesons are produced. (Photographs courtesy of Peter Fowler, University of Bristol.)

Figure 6.2. Energy spectra for some of the most abundant CR particles, hydrogen, helium, carbon, and iron. The general shapes of the spectra are similar though there are definite differences in slope in the high-energy region. Spectra below 100 MeV/nucleon are strongly influenced by the solar wind. (Reprinted with permission from *Opportunities in Cosmic-Ray Physics and Astrophysics,* copyright 1995 by the National Academy of Sciences; courtesy of the National Academy Press, Washington, D.C., and T. Gaisser.)

well-separated grains, produced by pions usually moving with speeds close to that of light. The number of pions, the multiplicity, increases with primary energy, with large fluctuations from one collision to another. At the same time the angular spread, the cone of angles encompassing the jet of pions, decreases as the energy increases. The opening angle of this cone provides a measure of the energy: a $1/20°$ cone corresponds to about 1,000 GeV. The emission of pions in these collisions takes place within about 10^{-23} sec of the initial impact, long before evaporation begins from the residual nucleus.

Charged pions, which produce the tracks seen in emulsions and cloud chambers, are accompanied by neutral pions, which do not produce visible tracks. Like all mesons, neutral pions are unstable, but their half-life is extremely short. After around 10^{-16} sec, they usually decay into two high-energy gamma rays, or (in just over 1 percent of the cases) directly into an electron-positron pair and a gamma ray photon. The gamma rays soon convert into more electron-positron pairs, and these in turn produce photons that convert into more pairs. The ensuing electron-positron-photon cascade builds up quite rapidly, then fades away as the energy is divided among too many particles for them to be able to reproduce further. In addition to the electron-photon cascade, another smaller cascade occurs that consists of protons, neutrons, and charged pions, all of which collide further and produce yet more mesons as long as they have sufficient energy.

Because a cascade develops differently in the atmosphere and in solid detectors, methods of estimating primary energy are based on different ways of sampling showers to determine the growth pattern. At one time, energy estimates were derived by laborious microscopic analysis of hundreds of tracks in individual collisions recorded in photographic emulsions. These studies led to considerable understanding of high-energy collisions many years before particle accelerators could approach such energies. Today we use calorimeters, devices that measure simultaneously the total energies of many particles rather than measuring energy particle by particle. An ionization calorimeter consists of several ionization detectors separated by slabs of material in which the shower can develop. The detectors thus sample each cascade's growth; when this growth pattern is compared with the predictions of the well-developed shower theory, one can calculate the total energy for each event. Like other electronic detector systems, the calorimeter enables many more events to be handled automatically than does the use of emulsions (no matter how picturesque that technique has been). The calorimeter technique is well suited to the energy range of 10^{12} to 10^{15} eV; at still higher energies the cascade is too extended to be confined within a calorimeter of any convenient laboratory size. Fortunately, the Earth's atmosphere again comes to the rescue.

After a high-energy collision in the upper atmosphere, a cascade develops until (after only about 10^{-4} sec) a swath of millions or even billions of particles arrives at sea level, spread out over an area that can extend for several square miles. In these extensive air showers (EAS) a concentrated core of particles is arrayed around the shower axis, with fewer particles at larger distances. The sheet of particles is a few meters thick in the center and perhaps 100 meters thick well away from the center. An array of counters, spread out on the ground in carefully arranged patterns, can detect only a small fraction of the shower particles but still enables researchers to reconstruct the shape and size of the shower and its energy. Thus even though only a handful of the highest-energy particles arrive each year—far too few for any conventional detector to capture—their multiplication in the atmosphere gives them away.

The discovery of air showers came accidentally in 1938 in France, where Pierre Auger and his colleagues were studying coincidences in a CR telescope and noted that they were recording coincidences between two counters that were several hundred meters apart. Further tests, some by Auger and others by George Rochester and J. G. Wilson at Manchester University using cloud chambers as well as counters, showed that large numbers of particles were arriving simultaneously, within the timing capabilities of their circuits.

Studying an EAS and identifying its various components requires a variety of detectors, some with lead shielding to filter out the soft component (the electrons and photons), others placed underground to look for the muons (from pion decay). Scintillators and Čerenkov counters are in widespread use, in arrays extending over large distances. Data are being obtained from an international web of systems such as those near Leeds, England; Volcano Ranch, New Mexico; Salt Lake City, Utah; Sydney, Australia; and Yakutsk, in

Siberia. The largest EAS array in use is the Akeno Giant Air Shower Array (AGASA), located about 80 miles west of Tokyo, near Akeno. AGASA is spread over an area of nearly 40 square miles and uses 111 scintillation detectors, each of 2 m², and 27 detectors for muons.

In an EAS, the charged pions that do not collide will decay (in about 200 millionths of a second) into charged muons, many of which reach detectors. Energetic muons constitute a durable component in the EAS; they behave a lot like electrons but are 200 times heavier. Unlike nucleons and charged pions, muons rarely interact with nuclei; they are 40,000 times less likely to produce the photons needed for shower growth. Consequently, the only important way a muon can lose energy in air is by gradually transferring small amounts of energy to electrons in the atmospheric atoms. Muons can thus arrive at ground level with much of their original high energies intact and then penetrate deep inside the Earth. Muon detecting systems have been operated down to depths of 12,000 feet below the Earth's surface in gold mines near Johannesburg in South Africa and in the Kolar gold field in India. Another deep counter array is located in a tunnel under Mont Blanc in the Alps. Primary protons with energies of around 10^{16} eV produce the muons that are seen in these counters; no other charged particles can penetrate to such great depths.

Because the shower's primary particle is not observed directly, its energy and nature can be deduced only from the products of collision, which multiply and then are detected. It is not possible to detect every particle in a shower, and the various systems can only sample a shower, with the detectors strategically placed. The precise mix of muons, electrons, and nucleons and their varying proportions across the shower front and at different heights in the atmosphere provide clues to the nature of the primary particle—but not without ambiguity.

EAS can also be produced by gamma ray photons with sufficiently high energy. Photon-initiated showers start with the primary photon converting into an electron-positron pair, and then the usual electron-photon cascade follows. If the primary projectile is a photon rather than a nucleon, the shower will contain very few pions, and although high-energy photons can fragment a nucleus, this is rare, so the muon and nucleonic components will be much reduced from their normal proportions. Showers that are muon-poor are likely candidates for having been produced by photons, but the problem is that muons might simply have missed the detectors rather than being truly absent.

Although photon primaries seem to be rare or absent among most regular EAS that come randomly from all directions, there is growing evidence that some showers are produced by photons coming from a very few astronomical objects such as Her X-1 and Cyg X-1. These objects, located in the constellations Hercules and Cygnus, are already known as emitters of X-rays and gamma rays and have been observed from scientific satellites such as the *Compton Gamma Ray Observatory (CGRO)*. The axes of their associated EAS point directly to them as they move across the sky. Photons of around 10^{12} eV have been identified with Cyg X-1, whose behavior at different wavelengths is quite irregular. This object is thought to be one of the better candidates for a black hole. A recent and exciting discovery is the detection of TeV gamma rays coming from the remnant of the supernova that was seen to explode in the year 1006. This object had already been identified as a source of X-rays, and the synchrotron spectrum indicates the presence of electrons that must have been accelerated to energies of hundreds of TeV.

Finally, we do not expect to see many EAS initiated by electrons. An electron with enough energy to produce a shower probably will not reach the Earth unless it has come from a relatively nearby source. During extended interstellar travel, it is far more probable that an electron's energy will have been significantly degraded through collisions with the many interstellar photons or through radiating energy in the synchrotron process in the interstellar magnetic field. This effectively places a limit on the maximum electron energy that we observe, a subject to which I return in Chapter 7.

Nevertheless, a Japanese group has detected 51 high-energy electrons during balloon flights in 1996 and 1998. Their detector system used a sandwich containing high sensitivity X-ray films, emulsions, and lead plates in which a shower can develop and be fully contained. They estimated the average initial primary electron energies at around 1 TeV and the distance of travel at no more than 1 kiloparsec.

To a fair approximation, there is little (but important) variation in *a*, the slope of the energy spectrum, around the value 2.8, for protons from 10^{10} eV on up to the highest energies. But it is just those small changes in a that carry important information on the way in which cosmic rays are accelerated and then travel in the galaxy. Different groups of researchers, with their different detector systems, disagree slightly on the precise values of a but the important features are clear. There is a region around 3–4 $\times$ 10^{15}, known as the knee, where the spectrum steepens from $a = 2.76$ to a value of $a = 3.2$. Then, around 10^{19} eV, there is another bend known as the ankle, where the spectrum flattens again to a smaller value of a. These changes are thought to reflect changes in the primary cosmic ray beam, in the proportions between protons

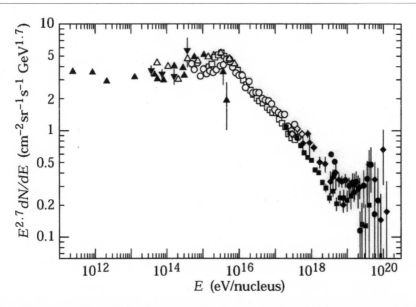

Figure 6.3. Spectrum for the highest-energy particles shown as function of total energy per nucleus. To accentuate differences in spectral form, the data points have been multiplied by $E^{2.5}$; a spectrum with a constant slope of -2.5 would then appear as a straight horizontal line. The persistent downward trend indicates that the true spectrum steepens with increasing energy. The bend at about 10^{16} eV is known as the "knee" of the spectrum. (Reprinted with permission from *Opportunities in Cosmic-Ray Physics and Astrophysics,* copyright 1995 by the National Academy of Sciences; courtesy of the National Academy Press, Washington, D.C. and T. Gaisser.)

and heavier particles and the way in which accelerating processes affect particles with different charges.

Below 10 GeV the situation is just as complex as with higher energies, but with differences. The energy spectrum bends over and displays a broad maximum around several hundred MeV, then decreases before beginning to rise again at the lowest energies observed, around 10 MeV. In addition, the low-energy spectrum varies significantly over the 11-year solar cycle, most noticeably at the lowest energies. This modulation of the energy spectrum is understood in only the most general of terms. Arriving cosmic rays have to traverse the solar wind as it blows out past the Earth, and they lose energy. A typical energy loss can be 400–600 MeV, so that a particle detected with 1,000 MeV would originally have been in the range 1,400–1,600 MeV before it encountered the solar wind. The residual energy is clearly related to the solar wind's strength at that time. We may simply not be able to detect any particles whose distant energies are well below 400 MeV. For this reason our knowledge of the unmodified shape of the energy spectrum in interstellar space is shaky below 1 GeV, yet this is potentially the energy region with most of the cosmic rays and most of the energy.

When we turn our attention to the energy spectrum of the heavy primary CR particles, those that are not protons, we need to take note of an inherent problem. Only the JACEE system, shown in Figure 6.4, allows direct identification (and then only to 10^{15} eV) of a primary particle through measurement of its ionization before the collision that generates the shower of debris, which is also observed. For higher-energy events, both the energy and any identification of the primary particle must come from the analysis of the shower structure and development. This analysis depends on theoretical extrapolation from measurements of lower-energy collisions under laboratory conditions, as well as on assumptions about the nature of the primary particle. Computer simulations are compared with the observations in identifying the primary particles and their energies. Showers initiated by protons, helium, and heavier nuclei develop differently, and the measurable quantities can be quite sensitive to the input to the calculations and the different computer models used.

A difference has been observed in the energy spectra of protons and helium nuclei, at energies up to 200 GeV. An international group of researchers detected over half a million protons and 30,000 helium nuclei during a long-duration balloon flight. They found a spectral slope of $a = 2.73$ for the protons and 2.63 for the helium. At higher energies and for heavier particles, the evidence is tantalizing but not yet definitive.

There appears to be a general agreement among a number of EAS groups

CHARGE
DETECTOR

TARGET

CALORIMETER

Figure 6.4. In the JACEE experiment, heavy primary particles can be identified from their tracks in the upper emulsion layers, and their energies calculated from the subsequent development of the showers. (Courtesy of John P. Wefel, Louisiana State University; *Astrophysical Journal* 502 [1998]: 278.)

that the proportion of protons and helium drops from the low-energy dominance of greater than 90 percent to around 50 percent by the knee, with the heavy particles, such as iron, making up about 12 percent (increased from far less than 1 percent). A 1999 paper resulting from a collaboration between groups at the Universities of Chicago, Michigan, and Utah reports that at 10^{16} eV, their data were in good agreement with calculations for events produced by primary iron nuclei. Support comes from the observations of the unique system that operated at a depth of 2.3 km in the Kolar Gold Field beneath the Himalayas. Muons produced in showers from multi-TeV primaries are the only charged particles that can penetrate to such a great depth.

To give some perspective to this picture, we can compare the energy of cosmic rays with that which we receive from the sun. Thus far, we have examined the energies of individual particles. When we multiply by the numbers of particles with different energies, we find the total energy that cosmic rays bring to the Earth every second. "Power" is the term used describe how rapidly energy is supplied or consumed; this can be measured in eV/sec, but ergs/sec is preferred. Ten million ergs per second constitutes 1 watt (1 horsepower is equivalent to 746 watts). The solar radiant power, or luminosity, L_o, is 3.8×10^{23} kilowatts. The rate at which solar energy reaches the Earth is about 1 kilowatt on each square meter.

If we add up all the energy carried by all of the CR particles, we find that the rate of arrival of CR energy on the Earth amounts to about 100,000 kilowatts (10^5 kw) or 2×10^{-10} kw/m^2—less than one billion times smaller than the energy arriving in sunlight, but comparable to the total energy that we receive in starlight. The implications of these numbers will be explored as we attempt to reconstruct the histories of the cosmic rays and consider their possible sources. But first, in the following chapter, we turn to the most energetic particles, which are at the center of one of the greatest puzzles today.

7

Exploration of the very highest range of CR energies rests entirely on the examination of EAS, as described in the preceding chapter. With this technique, and with research groups using detector arrays set out over larger and larger areas, the energy spectrum has been traced to quite remarkable energies. Even though there are several problems that remain to be solved in detail, there are very good reasons for feeling comfortable with the present overall picture: Most cosmic rays are produced and confined within our own galaxy. They originate in stellar regions with violent activity, such as supernova remnants, where they are accelerated and then wander around, perhaps with further acceleration, "filling" the galaxy, though with some leakage into the vast and almost empty regions of intergalactic space. Our qualified complacency rests on the chemical composition of the cosmic ray beam, the shape of the energy spectrum, and the overall energy picture, on plausible models for particle acceleration and on the identification of a number of sources of high-energy particles.

Within our galaxy, the average cosmic ray energy density is about $1-2$ eV per cubic centimeter, comparable to that of starlight. The frequency of supernova explosions and the magnitude of their energy release suffice to maintain this cosmic ray energy density inside the galaxy, but are grossly inadequate to fill the intergalactic space. We know of no sources of energy between galaxies, so that if cosmic rays "filled" intergalactic space to the same degree that they fill our galaxy, we would be faced with the presently unsolvable problem of finding an energy supply vastly greater than anything that we now know. And those sources would have to be hidden; they must not emit radiations that should already have allowed us to spot them using the traditional astronomical channels in visible and radio surveys.

And yet, there is growing and persuasive evidence pointing to some of the cosmic rays originating outside our galaxy. This evidence comes from the very highest-energy particles, those that we observe only through the secondary effects that they produce in our atmosphere.

The highest energy yet reported for an individual particle is 3×10^{20} eV, and this comes from an event recorded with an experimental approach used by the group at the University of Utah with their novel Fly's Eye Detector. The immensity of this particle's energy might be gauged from some comparisons. The average kinetic energy of a molecule in water at normal room temperature is in the region of 0.01 eV, and the total kinetic energy of all 3×10^{22} molecules in 1 cubic centimeter of water is roughly the same as that of a single CR particle with an energy of 10^{20} eV. If we could convert that one cosmic ray's kinetic energy to heat energy, it could boil that cubic centimeter of water, starting from just above freezing. Probably more topical: the energy of a 10^{20} eV particle is about the same as that of a baseball thrown at about 50 miles per hour

The paucity of these extraordinarily energetic particles makes this a topic where results are still few, though already compelling. In 7 years, the largest array of detectors, AGASA, recorded nearly 4,000 particles whose energies were over 3×10^{18} eV, 461 above 10^{19} eV, and only 6 above 10^{20} eV (see Figure 7.1). Other research groups have also reported observing particles above 10^{20} eV: four events at the Haverah Park array in England, 1 event at Yakutsk in Siberia, and one at Volcano Ranch in New Mexico. Energy estimates for individual events are considered to be good to ± 30 percent. Only one of these particles arrives on each square kilometer at the top of the Earth's atmosphere in each century! Observing these particles is an intimidating prospect, but the extent and the shape of the top end of the energy spectrum are sufficiently important that the infrequent arrival of these particles is well worth monitoring.

These particles are truly "ultrarelativisic"—their speeds differ from the speed of light only minutely, so that speed alone is a poor indicator of energies. A better indicator is γ (gamma), the ratio of total energy to rest-mass energy. For a particle with energy of 10^{20} eV, γ is about 10^{11}, and its speed differs from that of light by only a part in 10^{22}. For convenience, I will refer to particles with energies above 10^{17} eV as UHE (ultra-high-energy) particles.

Whereas the usual EAS system is designed to detect shower particles arriving at ground level, the Fly's Eye consists of an array of mirrors and sensitive phototubes that collect the faint light produced by the EAS particles on their way down through the atmosphere. The menbers of the Utah group seek out

Figure 7.1. Energy spectrum of CR particles observed with AGASA. The vertical scale has been multiplied by E^3. The numbers attached to each point indicate the number of events in each energy interval. The dashed curve represents the spectrum expected for extragalactic sources of CR particles distributed uniformly in the universe, with allowance made for the GZK effect. (Courtesy of M. Takeda, Institute for Cosmic Ray Research, University of Tokyo.)

light emitted by atmospheric nitrogen atoms that have been hit by shower particles and then fluoresce. They have compared their problem to that of detecting "a blue 5 watt light bulb streaking through the sky at the speed of light against a continuous backdrop of starlight, atmospheric airglow, and man-made light pollution. In addition, sporadic sources of lights such as lightning, auroras, airplane, and smokestack strobe lights . . . create a certain visual havoc" (*Nuclear Instruments and Methods,* p. 412).

Protons with energies up to 10^{18} eV can be contained within the galaxy. For such a particle, the curved path in the general galactic magnetic field will have a radius of 10^{21} cm, much smaller than the galactic diameter of 10^{23} cm. With increasing particle energy, the curvature of a CR path will approach and then

Figure 7.2 One of the Fly's Eye detector units, consisting of 256 hexagonal photomultipliers that gather the light. A glass filter removes background light and lets pass only the light from fluorescence of atmospheric nitrogen, produced by the passage of the cosmic ray particles. Each building on a Fly's Eye site is the size of a two-car garage, and many of the detector units are spread over two sites, to yields stereoscopic images of the developing fluorescent air shower.(Courtesy of John Matthews, University of Utah.)

exceed galactic dimensions, making it easier for such UHE particles to escape from the galaxy. The lower-energy particles are confined to the galaxy and arrive at earth isotropically, that is, uniformly from all directions. If higher and higher energy CR find it easier to escape, there will be a deficit in the numbers of particles arriving at earth from some directions, an anistropy that should increase with particle energy. There is evidence from the Fly's Eye for a small anisotropy for particles with energies between 2×10^{17} eV and 3.2×10^{18} eV; the AGASA team reports a 4 percent effect around 10^{18} eV, finding an excess of particles close to the galactic plane and correlated with the nearby spiral arms.

At energies above 10^{20} eV, the curvature of trajectory is comparable to the diameter of the galaxy, making it very difficult to confine these particles. From the computed reconstruction of a shower, the arrival direction of the primary UHE particle is considered to be accurate to within about 2°. With no effective magnetic field between galaxies, this means that the arrival direction of a UHE particle should point to within about 2° of its source or original direction. So far, the UHE particles show no clustering in arrival directions, such as pointing to known energetic objects such as galaxies, quasars, or active galactic nuclei. Also, if these particles had originated within our own galaxy in which the known matter is strongly concentrated toward the equatorial plane, we should be seeing evidence for an anistropy in arrival directions.

After so many surprises that have emerged in cosmic ray research why should we be surprised with yet one more, this time with the UHE particles? A still-significant problem was pointed out by Kenneth Greisen of Cornell University in 1966. In a paper entitled "End to the Cosmic-Ray Spectrum?" that appeared in the *Physical Review Letters*, he drew attention to an implication of the then-recently-discovered residual radiation from the cosmic fireball that accompanied the big bang, around 15 billion years ago. At almost the same time, G. T. Zatsepin and V. A. Kuz'min of the Lebedev Physics Institute arrived at the same conclusion as did Greisen, and their paper appeared in the Soviet *Journal for Theoretical and Experimental Physics*.

As these researchers pointed out, radiation from that gigantic Big Bang detonation has spread throughout the universe until by now its photons are everywhere, inside and between galaxies. The radiation has cooled from its initial billions of degrees until it is now at an effective temperature of close to 3 degrees above absolute zero (about 270 degrees Celsius below freezing). In the emptiness of intergalactic space, cosmic ray particles can collide with the photons of the cosmic background radiation (CBR, as it is termed) as well as with infrared photons from distant galaxies, and these collisions pose real threats to cosmic ray survival since there is nothing else to impede their passage. In contrast, within our galaxy, it is far more likely that the cosmic rays will collide with interstellar gas or dust and this limits the distances they can travel.

One might think that a collision between a cosmic ray particle with an energy of 10^{20} eV and a cosmic photon with an energy around 10^{-4} eV should leave the CR particle unscathed. (What wins when a 3-ton truck collides with a bicycle?) Things are very different when ultrarelativistic particles hit photons. Arthur Compton was awarded the Nobel prize in 1927 for his examination (while at Washington University in St. Louis) of collisions between X-ray photons and stationary particles (electrons). The related process (known as inverse Compton scattering) that is of concern here has high-speed electrons striking photons, and is also well understood. The theory of the inverse effect was first set out by two of Compton's younger colleagues at Washington University, Eugene Feenberg and Henry Primakoff, in 1948.

The analysis of Compton scattering and its inverse (involving electrons and photons) can also be applied to collisions between protons and photons. When the proton energy exceeds a threshold value, then instead of the target photons simply recoiling, an increasing fraction of the cosmic ray's energy goes into the creation of pions. The number of pions and the energy they absorb will increase very rapidly as the energy of the CR particle increases above that threshold energy. The net effect is that there should effectively be a cutoff in the primary CR energy spectrum—there should be no particles that reach us with energies above 10^{20} eV. This limit set by the inverse Compton effect, described by Greisen, Zatsepin, and Kuz'min is now called the GZK limit. Particles that might start with even higher energies should be degraded rapidly as they travel between galaxies, long before they can reach us.

Most cosmic rays have energies below 10^{10} eV, and protons dominate. As described in Chapter 6, the proportions change at higher energies, so that by 10^{16} eV, there is evidence for the presence of a significant proportion of heavy nuclei. It might therefore be thought that the problem with the UHEs could be avoided if the primary CR particles, each carrying more than 10^{20} eV, were heavy nuclei such as iron rather than protons. For ^{56}Fe, the most abundant iron isotope, there are 26 protons and 30 neutrons in each nucleus, and thus each nucleon has $(10^{20})/56$, or about 2×10^{18} eV/nucleon kinetic energy—far below the GZK cutoff for protons. This possibility was anticipated by the GZK authors, who pointed out that pion production by a large nucleus is disastrous for a CR heavy nucleus at much lower energies, and virtually none with energies of 10^{20} eV should survive.

We are left with a puzzle. There is no disagreement that the GZK cutoff poses a very real barrier to extended CR travel between galaxies, and it limits the distances that can be survived by UHE particles. However, several experimental groups, each with its own hardware, circuitry, and computers, have detected particles whose energies exceed that limit. As has been pointed out by the GZK authors, cosmic rays can travel in intergalactic space, but their lifetimes are limited to 10^8 years before their inverse Compton collisions wipe them out. We must therefore look for cosmic ray sources for the UHE particles within 10^8 light years of our galaxy. (In astronomical parlance, 1 parsec is 3.26 light years, so 10^8 light years is about 30 Megaparsecs, or Mpc. Recent calculations have raised this limiting distance to about 50 Mpc.)

What sort of sources might one look for? It would be reasonable to expect that a source of UHE cosmic rays should also show other signs of violent activity. Surveys by the Energetic Gamma Ray Telescope (EGRET) on the *Compton Gamma Ray Observatory (CGRO)* have yielded no candidate, and a similar blank has been drawn by surveys of extragalactic radio sources with small redshifts, indicating they are within the GZK limit of a few tens of Mpc. (Radio emission can be indicative of synchrotron radiation, where the radiating electrons come from pion production in CR collisions.) The group at the Smithsonian Astrophysical Observatory has searched for TeV gamma rays around the directions of the 10^{20} eV particles, but found nothing.

A similar blank was drawn by scientists of the AGASA collaboration who examined the arrival directions of their 10^{20} eV air showers, gamma ray sources detected by EGRET, and extragalactic radio sources with redshifts less than 0.02. These modest redshifts indicate galaxies that are our neighbors (relatively, on the cosmic scale of distances). However, a few scientists have suggested a correlation between the UHE directions and some quasars. The directional alignments are only within about 10 degrees and the quasars are at distances vastly greater than allowed by the GZK effect. If these directional associations are indeed correct, then we would have to look for exotic mechanisms, never before observed. One possibility is that, in those distant quasars, violent processes produce as-yet-undiscovered particles that travel toward us over much of those great distances, then, within the GZK range, transform into the well-known cosmic ray particles that we detect. Other exotic particles or processes have been considered, and the field is open to considerable theoretical ingenuity. Before we reject any of these out of hand, the observable consequences of each model need to be considered—would they suggest objects or radiations that we can exclude on the basis of present surveys, or perhaps have not even looked for?

A rather different identification has been entertained for some of the highest-energy primaries: high-speed dust grains. Huge numbers of microscopic dust particles pervade interstellar regions like a celestial smog, absorbing light preferentially at the shorter (blue) wavelengths and making distant stars appear redder. A few of these grains, it has been suggested, might be accelerated and could, on colliding with atoms in the atmosphere, produce EAS. Although this hypothesis is sometimes used as a possible explanation for the highest-energy EAS primaries, the best identification still seems to be the conventional protons, helium and heavier nuclei.

Detection of the few UHE particles has raised important questions whose answers require far more extensive observation. A gigantic EAS array, to be known as the Pierre Auger Southern Observatory, is being constructed in Argentina, with completion planned for 2001. Brazil, Britain, France, Mexico, and the United States are collaborating in the construction of the particle detectors for the array, and Germany and Italy are providing telescopes that will be used for the simultaneous detection of the light that the EAS particles produce as they race down through the atmosphere. There will be 1,600 particle detectors, spaced 1.5 km apart, over an area of 3000 km². If sufficient funding can be obtained, a twin system will be constructed in Utah, so that there will be one in each hemisphere.

A very different approach to the detection of the rarest particles is the *OWL* project now being planned. The *Orbiting Wide-Angle Light* collector is to consist of two earth satellites that always look down, to detect the flashes of ultraviolet fluorescence that UHE CR particles produce in the atmosphere. Each satellite will have 10 m² of detector, will view 1 million square kilometers of atmosphere, and should detect 500–1,000 events each year.

What we see here, once again, is the raising of a tantalizing question through cosmic ray research. What is first needed is the accumulation of much more data from AGASA and the Auger Observatory, perhaps augmented by data from *OWL*.

In the next chapter, we will begin to weave together a broad canvas of the cosmic ray picture by examining the clues provided by the nuclei, and in the following chapter we will review ideas on the origins of our cosmic radiation.

8

NUCLEAR CLUES

RACKING DOWN THE ORIGIN of cosmic rays takes us along a path that, over the years, has followed many twists and turns and has not yet been completed. As we have seen in Chapter 7, there is strong evidence that particles with the very highest energies must come from outside our galaxy, but these represent only a minuscule fraction of the CR, in both numbers and total energy. The vast majority of the CR originate and spend their entire lives within our own galaxy. In this chapter, we will see how nuclear physics provides clues that set strong constraints on the types of models we can entertain. Then, in the following chapter, we will put this all together into a comprehensive picture.

We have accumulated a great deal of information about many aspects of cosmic rays. We know the major characteristics of the solar energetic particles and the solar wind's effects on the galactic CR. The relative abundances of nuclei have been well charted, in detail up to atomic numbers around 60, and major features have been determined for the rest of the periodic table. There have been impressive improvements in the identification of individual isotopes. The energy spectra of the more abundant species have been measured for the energy range that encompasses more than 95 percent of the total CR energy. How can all this information be pulled together so that we can start to construct a coherent picture of cosmic rays?

In a popular view of science, further progress comes through induction: we assemble an array of well-established specific facts and from them, by compelling logic, construct a general theory, broad in its sweep. Some scientific problems readily yield to this form of attack, but many do not. Instead, a more fruitful approach is what Karl Popper has termed conjecture and refutation. In this widely followed procedure, a conjecture (or model) is suggested and its consequences explored by logic, deduction, and computation. The confrontation between predicted consequences and actual observations can either lend support to the model or require its modification or even rejection.

Where does the conjecture or model come from? There is no simple or unique answer. Models can come from older models, now creaking under the weight of newer experimental results. They can come from hunches, flashes of inspiration, usually based on experience or on comparison with problems that seem similar. They emerge after the exhaustion of alternatives. "They can be reached only by intuition, and this intuition is based on an intellectual love of the objects of experience," as Einstein once said (*Imagination and the Growth of Science,* p. 9).

The model for most cosmic rays that is now generally accepted—at least as a starting point—locates their origin and confinement inside our own Milky Way galaxy. The initial charge and energy spectra of the primary particles are assumed, and their subsequent propagation is studied. The particle energies are tracked through any further acceleration or energy losses. Some cosmic rays escape from the galaxy; others have collisions that produce secondary fragments. Changes in the relative abundances of different particles are tracked through collisions that remove some of the primaries and create secondaries such as the L nuclei, lithium, beryllium, and boron. The energy spectra and relative abundances are part of the input, along with the various mean free paths and cross sections.

Finally, calculations yield predictions for what we can expect to observe at Earth. These computations include many physical quantities (parameters); some of their values can be measured in the laboratory while others must be estimated and later adjusted to agree with observations. The more adjustable parameters a model contains, the closer we can come to "explaining" our observations; on the other hand, we might have less confidence in such a model because different sets of parameter values might yield equally good fits. A good theory keeps its adjustable parameters to a minimum.

The composition, or relative abundances, of cosmic rays clearly provides a logical starting point for model building. We can start by comparing the relative abundances of CR nuclei to abundances in the solar system, the material most accessible to us. We find a broad similarity and a striking difference. The similarity shows in a steady trend toward lower abundances as we proceed from the lightest nuclei, hydrogen and helium, to the heaviest. The immediately obvious difference lies in the abundance of the L nuclei, which are quite plentiful in the CR beam but about 100,000 times less abundant in the solar system. What does this tell us, and why are there so few L nuclei at large? The answer comes from one of the great success stories in astrophysics, the theory of nucleosynthesis.

Our best model for the evolution of the universe starts with an extraordinary explosion, the big bang, when all the contents of the universe were jammed into an unimaginably small volume. In those first few high-temperature minutes, most of the helium in the universe was created by hydrogen fusion. Although helium is still being formed inside stars today, this source is not sufficient to produce the known amount of helium. Recognition of the cosmological production of helium during the big bang solved this long-standing problem. Deuterium was also produced in the big bang, and the proportions of deuterium and helium today provide clues to the conditions then, about 15 billion years ago. Expansion of the universe, accompanied by a drop in the temperature, later permitted protons to combine with electrons to form hydrogen atoms.

Much later the dispersing material began to clump into galaxies, and after a hundred million years or so the first stars began to form, composed mostly of hydrogen. The time scale for further development depends on the mass of each star, but the initial behavior is qualitatively similar for all stars. The gravitational attraction between the atoms that make a star produces a high pressure at the star's center. Hydrogen subjected to such a high pressure will be at a very high temperature. In the sun this central temperature is about 15 million degrees; once again the hydrogen is totally ionized into protons and electrons. At such high temperatures, colliding protons can fuse to create deuterons. Further collisions and a web of nuclear reactions lead to the production of helium nuclei, with the release of considerable quantities of energy. This energy, much of it in the form of high-energy photons, percolates slowly to the surface, from which it is radiated outward, with less than one billionth arriving on Earth to keep us warm.

In each chain of fusion reactions, four hydrogen nuclei combine to form a single helium nucleus. The total number of nuclei is therefore reduced—one helium nucleus exists where before there were four protons, and a helium nucleus is not much larger than a proton. With this reduction in volume, the star's central core cannot withstand the pressure of the overlying material and contracts, raising its central temperature. When the central temperature reaches 100 million degrees, helium nuclei fuse in further reactions and carbon and oxygen nuclei begin to form (out of three and four helium nuclei, respectively). For a star of solar size, this stage is reached after around 10 billion years and is accompanied by a general expansion, into the red giant phase.

For more massive stars, the pace of evolution quickens and nuclei heavier than oxygen nuclei can be formed. Some reactions release neutrons, and these can be captured by other nuclei. In many stars the buildup of heavier nuclei occurs by this steady accumulation of neutrons. Some new nuclei are radioactive and decay before another neutron can be captured, but the accretion continues through successive neutron captures. This slow (or s) process ultimately produces nuclei as large as lead ($Z = 82$) and bismuth ($Z = 83$), the two heaviest nonradioactive nuclei.

An interesting feature of these nuclear reactions is that L nuclei are created but just as rapidly are consumed in further nuclear reactions. The general rarity of these three L elements (lithium, beryllium, and boron) can now be seen as a natural consequence of nuclear properties and the reactions taking place in the interiors of stars. In contrast, the only viable theory for the presence of the relatively large numbers of L nuclei among cosmic rays is based on the fragmentation that some of the heavier CR particles undergo during interstellar travel (see Figure 8.1). The L nuclei are thus important as indicators of the distances traveled by cosmic rays and the resulting numbers of collisions.

Figure 8.1. Fragmentation of a cosmic ray magnesium nucleus (charge $Z = 12$) into six high-speed alpha particles, which emerge in the tight bundle of tracks. The more widely spread black tracks are fragments of the target nucleus in the emulsion. Collisions of this sort, between primary cosmic ray particles and interstellar atoms, provide a means by which we can estimate the distance and time of travel of the cosmic rays. They may also be the major source of lithium, beryllium, and boron found on Earth. (Photograph courtesy of Peter Fowler, University of Bristol.)

Moreover, it appears that almost all of the Earth's content of these three light elements was produced by interstellar CR collisions.

In some massive stars the cycle of central fuel consumption, subsequent collapse, and further heating becomes unstable, and as a result many more neutrons are generated during an explosive stage. The torrent of neutrons (in this rapid, or r, process) can be so large that many neutrons are captured and nuclei heavier than lead or bismuth are produced before radioactive decay shuts off this growth. These heavy nuclei are all radioactive, and most have short half-lives. Only thorium ($Z = 90$) and uranium ($Z = 92$) have isotopes with long enough half-lives that they have survived the 4.5 billion years of the solar system's existence. Their presence on Earth proves that the solar system formed from interstellar material that had already been enriched through the r process. Short-lived nuclei with atomic numbers between 84 and 89 and beyond 93 do not occur in nature, but during the past 40 years they have been synthesized in large accelerators and also in test nuclear explosions.

The explosive r-process stage of a star's history occurs during an event that appears as a supernova—the extreme, sudden, and unpredictable brightening of a star, as happened so spectacularly in the Large Magellanic Cloud in February 1987. In such an explosion, newly created heavy nuclei are hurled out into the interstellar medium. Some drift into regions of higher density, where they are among the atoms condensing to form the next generation of stars.

The theory of hydrogen fusion reactions in the sun is based largely on the pioneering work of Hans Bethe of Cornell University, who won the Nobel Prize in 1967. For heavier nuclei, the classic research was carried out by Geoffrey Burbidge, Margaret Burbidge, Willy Fowler, and Fred Hoyle of Cal Tech and Cambridge University. Their 1957 paper (usually referred to as B²FH) set out the roles of the r and s processes. (See Figure 8.2). Quite independently, Al Cameron, in Canada, derived similar results at the same time. Of course, the theories of stellar evolution and nucleosynthesis are far more complex than the brief summary I have given here, and other processes in addition to the r and s processes come into play, though in less prominent roles.

The properties of different nuclei and their behavior in nuclear reactions can often be studied in the laboratory, but sometimes formulas have had to bridge the gaps between lab and cosmos. With computers, one can model a stellar interior or a star's evolution during the catastrophic supernova stage. Some of this analysis can be used when researchers tabulate observed abundances, element by element and isotope by isotope, by looking at materials from the Earth, moon, and meteorites as well as data from astronomical spectral analysis. Solar system material does not come from only the r or the s process, but appears to consist of a mixture, with additions from other nucleosynthetic reactions.

Another outcome of the r-process calculation is a prediction of the abundances of the long-lived isotopes ^{232}Th, ^{235}U, and ^{238}U, much heavier nuclei than can come from the s process. Thorium has a half-life of 14 billion years, and after ten successive but unevenly spaced radioactive decays it ends with a stable isotope of lead, ^{208}Pb. The uranium isotopes, with half-lives of 7×10^8

Figure 8.2. Margaret Burbidge, Willie Fowler, Fred Hoyle, and Geoff Burbidge, the authors of the B²FH landmark paper on stellar nucleosynthesis, gather to celebrate Fowler's seventieth birthday and the twenty-fourth anniversary of that paper. (Photograph courtesy of Margaret Burbidge.)

and 4.5×10^9 years, produce the lead isotopes ^{207}Pb and ^{206}Pb, respectively. Knowledge of the half-lives and abundances of all these isotopes has enabled researchers to estimate the age of the solar system at 4.6 billion years. This result provides a calibration for the time scale for stellar evolution, and from it we can infer the age of our galaxy to be 10 to 20 billion years.

The importance of these studies for a cosmic ray model is that there may be signatures in the CR charge spectrum that tell us about their sources. If, for example, cosmic rays come from supernovas with their r process, then we would expect to detect some transuranic nuclei. Other clues to the roles of the r and s processes, in addition to the presence or absence of the heaviest nuclei, are the proportions of intermediate nuclei among cosmic rays.

The approach to CR modeling should now be coming into clearer focus. We can assume that CR particles start from a certain type of stellar source, and we can further assume a set of relative abundances (such as from r or s processes) at the time of origin. We can then trace the history of these particles as they travel through the interstellar medium, and the L nuclei can be used as the first distance indicator. In carrying out these calculations, we must allow for the ways in which fragmentation and energy loss rate depend on particle energy.

With the background just described, we are ready to assemble and assess a comprehensive model for cosmic radiation. After allowing for effects of the geomagnetic field, we find that, in general, galactic CR arrive at the top of the Earth's atmosphere uniformly from all directions. This isotropy can mean either that the particles do not come from any one region or that they have been deflected so often in their travels that they have lost all sense of their original directions of travel. Two pieces of astronomical evidence are relevant to the puzzle.

Most cosmic rays do not come from the sun, so it is plausible to look for sources of galactic CR among stars that are more luminous than the sun or more energetic in some other way. But stars are definitely not distributed at random; rather, they are concentrated in the equatorial region of the galaxy, along the spiral arms in the Milky Way. If cosmic rays traveled along straight lines from stellar sources to Earth, we should see more of them arriving from the galactic disk, but that is not what we observe. Thus we must look for alternative explanations for the isotropy.

Across the enormous interstellar distances there is a general magnetic field, well organized in some regions and more tangled in others. The strength of this field, deduced from astronomical measurements, is about 10^5 times weaker than the magnetic field on the surface of the Earth. On the average, the energy density of this interstellar magnetic field is around 1 eV/cm^3, comparable to the energy density of starlight. Another indication of the presence of the interstellar magnetic field comes from radio astronomy. Radio signals have been detected from many regions that are remote from clearly identified objects such as stars and galaxies. The spectra of these radio signals unambiguously identify their origin in synchrotron radiation, in which high-speed electrons travel along curved paths in a magnetic field and emit radio waves. This cosmic radio noise (as it is sometimes called) leads to an estimate of the magnetic field strength that agrees well with that obtained from optical measurements. Another measurement of the interstellar magnetic field comes from pulsar radio signals, and these quite different methods yield general agreement.

Because the cosmic rays are electrically charged particles, they are forced to travel along curved paths in the interstellar magnetic field. In places where the field's irregularities are moving, the cosmic rays may be accelerated or slowed down in addition to being deflected. In general, we expect the CR paths to be severely tangled.

When a charged particle in a magnetic field moves around a circular arc, the radius of that circle depends on the field strength, B, as well as on the particle's energy, E, and charge, Z. For high-energy particles, the formula that expresses this relation (in units that are most appropriate for CR and interstellar conditions) is:

$$R = 3.3 \times 10^{12} E/(ZB),$$

where the radius (R) is in centimeters. We encountered this behavior of charged particles in Chapter 3 when examining the effects of the Earth's magnetic field. On the galactic scale the field strength is much weaker, and the size of the region in which particles travel is far larger than the solar system. Table 8.1 lists the radii of curvature of the paths of CR particles having typical energies for an interstellar field of 3 microgauss.

These numbers tell us that, for example, a 1-GeV proton will be moving along a circular arc of radius 1.1×10^{12} cm in a typical interstellar field. This dimension is less than one tenth of the Earth-sun distance, so the 1-GeV particle will have its path coiled up on a scale that is small compared to the solar system and minuscule in the galaxy, where distances are millions of times larger. If the magnetic field extends over a region much larger than that radius, then a CR proton can be confined within the region. In contrast, if the extent of the magnetic field is much smaller than that radius, the proton will not be confined, even though for a short while it will follow that circular path. At the high-energy extreme, the radius of curvature for a particle with an energy of 10^{20} eV is comparable to the diameter of our galaxy and far larger than the thickness of the galactic disk. Thus the EAS primaries with such high energies cannot be effectively contained within our galaxy, and this immediately introduces the possibility of an extragalactic origin, at least for the highest-energy particles—a subject already discussed in Chapter 7.

Because the paths of cosmic rays are so tangled, the important thing to know is the actual distance traveled. This is like measuring the length of a knotted piece of string after straightening it out rather than using the diameter of the knot. As cosmic rays travel, they occasionally encounter atoms in the interstellar medium, and the number of collisions (with resulting fragmenta-

Table 8.1 Radii of curvature of proton paths
in a 3-microgauss field.

Proton energy (eV)	Radius (cm)
10^9	1.1×10^{12}
10^{11}	1.1×10^{14}
10^{15}	1.1×10^{18}
10^{17}	1.1×10^{20}
10^{19}	1.1×10^{22}

tion or creation of mesons) is proportional to the total number of atoms along the path (Figure 8.3). If the interstellar medium had the same density (number of atoms per cubic centimeter) everywhere, we could simply convert the path length in centimeters to the equivalent number of atoms encountered. Where the density is irregular, we can still figure out how many atoms are along the path if we allow for the variations in density.

To introduce a familiar analogy, the behavior of CR particles could be compared to the experience of running along the concourse in an airport. The distance you actually run will certainly be longer than the direct distance to the gate, and the number of your near collisions will depend on the density (numbers) of passengers, as well as their sizes and speeds. You will have fewer collisions when you run with the stream of other departing passengers than when you run against a stream of those arriving. The total distance you run must be measured along your zigzag path, or it could be expressed in terms of the number of near collisions.

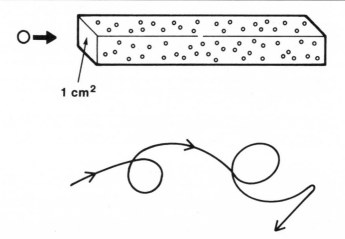

Figure 8.3. Cosmic ray path lengths can be expressed in terms of the number or total mass of atoms encountered. Along a straight path, the number of atoms within an area of 1 cm^2 provides this measure, usually given in gm/cm^2. Along an actual path, much bent by the interstellar magnetic field, the equivalent straight length is obtained by adding up the contributions from the many small and successive, relatively straight segments.

With CR travel it is customary, and more useful physically, to express distances by the total mass of all atoms encountered, and to do so in units of grams per square centimeter. Imagine the path straightened out (Figure 8.3) and a particle traveling a distance D cm through a region containing n atoms/cm^3, with each atom having a mass m grams. Then a total of nmD grams (of interstellar matter) will be within a volume that is D cm long and 1 cm^2 in cross section. For example, in the interstellar medium, a typical density is 1 hydrogen atom/cm^3, and each atom has a mass of 1.6×10^{-24} gram. A CR path of 10^{21} cm can then be expressed as

$$(1 \text{ atom/cm}^3) \times (1.6 \times 10^{-24} \text{ g/atom}) \times (10^{21} \text{ cm}) = 1.6 \times 10^{-3} \text{ g/cm}^2.$$

If the density of the Milky Way galaxy were everywhere 1 atom/cm^3, its diameter of 10^{23} cm would present a target of 0.16 g/cm^2. How many collisions a cosmic ray actually undergoes also depends on the sizes (cross-sectional areas) of the cosmic ray and the interstellar atoms, but the g/cm^2 figure provides a general index to CR travel history.

These calculations can also be applied to the propagation of carbon and oxygen nuclei, which are plentiful in prospective CR sources, in contrast to the L nuclei—Li, Be, and B—which are very rare. The L nuclei that we detect among the cosmic rays at Earth must be fragmentation products from interstellar collisions of cosmic ray C and O nuclei. We can calculate that C and O nuclei must have traveled along paths of around 5 g/cm^2 to produce the observed numbers of L nuclei. The distance is somewhat dependent on C and O energies, being about 7 g/cm^2 for 1 GeV particles and 2 g/cm^2 for 10 GeV particles. For an average density of 1 atom/cm^3 in the interstellar medium, the 5 g/cm^2 path corresponds to a distance of $5/(1.6 \times 10^{-24}) = 3 \times 10^{24}$ cm, or 30 times the galactic diameter. The CR paths must indeed be tangled; from the L nuclei abundance we deduce that a typical C or O particle has traveled a distance of about 30 times the galactic diameter between its origin and arrival at Earth.

We can also calculate the time taken for this long journey. For a particle of 1 GeV, the speed is close to nine-tenths the speed of light, or 2.7×10^{10} cm/sec. The 5 g/cm^2 will be covered in 3×10^{24} cm/2.7×10^{10} cm/sec $= 10^{14}$ sec, or about 3 million years. How reliable is this estimate of the travel time? It is based on distance (expressed in g/cm^2), speed, and interstellar density. The g/cm^2 value is deduced from the abundance of the L nuclei (the values are slightly different when one is using iron nuclei and their corresponding fragmentation products), and this figure is probably accurate to within about 50 percent. The speed is well known, but the value assumed for density (1 atom/cm^3) is at best an initial guide. Since density varies greatly across the galaxy, it is necessary to introduce a calculation quite different from any we have seen thus far to determine the density of the region(s) where cosmic rays travel.

Most nuclei assumed to be injected at CR sources are stable (nonradioactive). Among the fragmentation products generated during their subsequent travel, however, there are many radioactive nuclei. If we start with an assembly of 1,000 radioactive atoms, then after one half-life, half of the original nuclei will remain and the others will have decayed into lighter nuclei, after emitting alpha particles or electrons. After another half-life 50 percent of the remaining nuclei will have decayed, so that only 25 percent have survived; after another half-life the survivors are down to 12.5 percent, and so on. The decays of radioactive nuclei start immediately after creation, and there is no way we can identify which specific nuclei will decay promptly and which will hold out for many half-lives. All we can do is describe the process in statistical terms, which turns out to be sufficiently accurate.

Among the beryllium isotopes produced by the fragmentation of heavier CR nuclei, only ^{9}Be is stable, but ^{10}Be has a half-life of 1.6 million years—comparable to the travel time of cosmic rays. If the travel time were much longer than 1.6 million years, then most of the ^{10}Be nuclei would have time to decay, and we would not detect many among the CR particles. Conversely, if the travel time were much shorter, we would expect most ^{10}Be particles to survive and to show up when the Be isotopes were identified. Our best measurements, from the *Ulysses* and *Voyager* spacecraft, show that 4.6 percent of the beryllium nuclei are ^{10}Be. From this and from our knowledge of how C and O nuclei fragment, we can calculate that their average travel time (for particles around 100 MeV/nucleon) is 18 ± 3 million years. A similar calculation can be performed for an isotope of aluminum, ^{26}Al, which has a half-life of 700,000 years and is a fragmentation product of silicon and calcium. This travel time is found to be 16 ± 3 million years. For the chlorine isotope, ^{36}Cl with a half-life of 300,000 years, the travel time is found to be 11 ± 4 million years. With these estimates of the travel time, we can refine the value for the average density of regions where cosmic rays travel to 0.2–0.3 atoms/cm^3. This is an important result because it is significantly less than the general average of 1 atom/cm^3 in the galactic disk, and implies that a considerable fraction of the cosmic rays' travel must be in the galactic halo.

Another product of these calculations is a set of abundances for cosmic rays at their sources, before interstellar fragmentation has begun. For example, the relative source abundances of carbon:nitrogen:oxygen of 1:0.12:1.2 give the best fit to the ratios observed near Earth (1:0.023:0.9). Similar analyses have been carried out for other nuclei, such as those with charges 21–25, many of which are fragmentation products from primary iron nuclei ($Z = 26$).

I should add a word of explanation here. The calculation just outlined is typical of many in astrophysics. For many processes, the relevant quantities cannot be pinned down with the precision expected under laboratory conditions; agreement to within 10 or 20 percent is often considered good (or fortuitous). Knowing a quantity to within a factor of two sometimes suffices to exclude some models and concentrate the attention of researchers on others.

Calculating the effects of propagation requires extensive computer codes, starting with an equation for each type of particle that presents all relevant quantities and processes. The gradual change in composition of a CR beam is tracked by computing the cumulative effects of successive changes during small increases in the distance traveled. Allowance is built in for collisions, fragmentation, energy losses by ionization and radiation, the creation or injection of some particles, and the decay of others. A set of interlocking equations is needed because fragmentation that destroys a nucleus of one type (such as C or O) will produce nuclei of other types (such as Li, Be, or B).

By now we know with reasonable precision the values for most terms in these equations; some terms may be negligible in special cases, such as when particles with energies over 10 GeV are considered. Moreover, we can investigate what happens when starting assumptions are changed; for example, we can assume different starting proportions of C, N, and O and examine the resulting production of L nuclei and ^{10}Be to obtain the best agreement with our measurements.

In carrying through these calculations, we must also make assumptions regarding the regions in which cosmic rays are moving. Is the travel entirely in the galactic disk, or is some part of it in the less dense halo? The currently popular model is that of a "leaky box," in which cosmic rays are mostly confined to the disk by the galactic (interstellar) magnetic field. Because the energy density of cosmic rays is comparable to that in the magnetic field, they exert a major influence on the field, which bulges out of the disk into the halo. Leakage of cosmic rays into the halo depends on their energy; high-energy particles have curved paths with larger radii of curvature and thus are more likely to get into the halo than less energetic particles with more tightly coiled paths.

The distances that particles travel depend on their energies. We see this when we examine the energy dependence of the abundances of fragmentation products such as the L nuclei. An extensive analysis shows that at any given energy there is a distribution of path lengths. Some particles travel considerably greater distances than others, and it has turned out that among heavier particles with high energies there is a shortage of short path lengths. Shorter path lengths produce fewer collisions and thus affect the production of secondaries.

In much of this chapter I have concentrated on nuclei in cosmic radiation and how their abundances reveal their propagation histories. I have neglected CR electrons, whose near-Earth presence can be observed directly and whose distant presence is detected by the radiation they emit. This is an important topic that I will return to in Chapter 9, but first there is more to say about the nuclear component and its origins.

9

THE ORIGIN OF COSMIC RAYS

OVER THE YEARS, as our fund of information has grown, various models of the origin of cosmic rays have been entertained, supported, and sometimes neglected—or recycled. No model yet satisfactorily encompasses all of our knowledge. The changing views of partisans in this long-running debate provide a fascinating insight into the swings of fashion and consensus. What I outline here is the model involving supernovas that now has wide support; it provides a framework within which to plan further observations and to judge the significance of each piece of evidence.

Supernovas are now generally considered to be the best candidates for the sources of the energy and some of the CR particles. The evidence regarding total energy is strongly persuasive, especially in the absence of any serious competitor. Some events that we now call supernovas were noted long ago, before there was any understanding of their true nature. The apparent permanence of the stars was deeply ingrained in ancient cosmologies. Occasionally, however, a star would suddenly appear where no star had previously been seen. This new star (called *nova stella* in medieval times) would brighten rapidly, then gradually fade. It was not until 1934 that the name "supernova" was used, by the Mount Wilson astronomers Walter Baade and Fritz Zwicky, to describe novas whose exceptional brilliance set them apart from all other novas. Baade and Zwicky suggested that supernovas were the sources of cosmic rays, although they thought that CR were mostly positrons coming primarily from supernovas in other galaxies.

The initial classifications of historically recorded events as supernovas rested on brightness alone, but other characteristics show them to be very different in their nature from the less luminous novas. Both novas and supernovas have been seen in our galaxy, as well as in many external galaxies. The relatively frequent novas have sometimes been regarded as CR sources but are not currently favored. By now, it is clear that some, perhaps most, of the CR come from supernovas that seem to be the only sources of the large amounts of energy needed. There are, though, questions that remain: do the CR get their energy promptly from the SN explosions or only later? Answer—some of both! In this chapter, I lead us through these problems and our current ideas for solutions.

No supernova has been seen in our galaxy since 1604, when Johannes Kepler and many other astronomers in western Europe witnessed the spectacular outburst of a new star in the constellation Serpens, not far from the positions of Mars and Jupiter at that time. At a place where no star was visible on October 8, a bright new star was first seen two nights later. It continued brightening until early November and then faded slowly, but it was still visible a year later. Records of extensive observations of the star's brightness, made by astronomers in Europe, China, and Korea, are in good agreement. At its maximum, SN (supernova) 1604 was as bright as Venus. Among the other historic supernovas are those seen in 1006, 1054, and 1572, the last one described in great detail by Tycho Brahe, the famous Danish astronomer.

All these early supernovas preceded Galileo's invention of the astronomical telescope in 1609. Astronomers did not have the opportunity to examine a really close supernova with modern telescopes and spectroscopes until the outburst in February 1987 in the Large Magellanic Cloud, a relatively nearby galaxy. Unfortunately, this object is in a part of the sky where it cannot be seen by the major northern hemisphere observatories, optical or radio. Nevertheless, it is being followed over the entire e-m spectrum, and a special expedition went to Australia to carry out gamma ray observations with balloons. The measurements made so far give strong support to the standard theory of nucleosynthesis in supernovas. The steady decline of the visual brightness of SN 1987A suggests that the source of this light is a radioactive isotope of cobalt, ^{56}Co, produced by the decay of the nickel isotope, ^{56}Ni, that was synthesized in the explosion. Satellite observations have detected the expected high-energy X-ray photons.

Astronomers are always on the alert for new supernova explosions, especially in our galaxy, where they may be close enough for detailed examination. There are many that occur in other galaxies and notification of a discovery is sent out promptly. Historical records of astronomical events have proven useful for many purposes, including the study of supernovas. Records from European countries tend to be spotty, but those from China, Japan, and Korea are often extensive. China had court astronomers almost continuously for 2,000 years, until 1911, and they made copious notes of observations of the heavens with special mention of changes. These records have been scoured by modern astronomers eager to identify supernovas. Transient events were noted for astrological purposes; appearances of what the Chinese called "guest stars" were of particular interest. These are now recognized as including meteors, comets (with and without tails), and novas, as well as supernovas. The completeness of Chinese records is exemplified by the inclusion of every return but one of Halley's comet since 240 B.C.; in addition, the supernovas of 1006, 1054, 1572, and 1604 are all well documented.

Optical, radio, and X-ray telescopes have all scanned those parts of the sky in which putative supernovas were noted, and the astronomy of supernova remnants (SNR) has developed into an active field. In all of the clearly identified SNR, radio astronomers have detected synchrotron radiation emitted by energetic electrons as they spiral in a magnetic field. The electrons are usually confined in a roughly spherical shell, containing material expelled at high speed from the site of the original stellar explosion. This clear indication of high-speed electrons is one reason for considering supernovas as possible CR sources.

Until the appearance of SN 1987A, by far the best studied of all of the SNR was the Crab nebula, located at a distance of 2 kiloparsec (6,500 light years) from Earth. For many years the Crab was little more than a curiosity. Charles

Messier listed it as the first object in his 1771 catalog, designed to identify hazy objects that astronomers should ignore in their search for the comets that were then fashionable. (This was soon after the first return of the comet that Halley had so spectacularly predicted for 1758.) The Crab is now also known by its catalog number, M1. Interest in the Crab increased markedly after its identification in 1948 as the first galactic emitter of both radio and visible radiation. In the first radio survey of the sky, the Crab was among the brightest objects listed. From analyses of the shape of the Crab's spectrum and its polarization, it became clear by 1955 that much of the Crab's radiation at visible wavelengths was synchrotron radiation, produced by electrons with energies above 10^{11} eV.

The synchrotron character of some of the Crab's radiation had been predicted by the Russian theorist Iosef Shklovsky in 1953 (in a paper in a Russian journal that was not then widely read in the West). In addition, traditional astronomers had not yet had to deal with synchrotron radiation, a topic more familiar to high-energy particle physicists. As Shklovsky wryly noted in his 1960 book, "In September 1955, the proprietors of the greatest telescope in the world finally expressed active interest in the problem of the polarization of the Crab" (p. 308). Then Walter Baade, using the new 200-inch telescope on Mount Palomar, discovered the Crab's polarization. The pictures shown in Figure 9.1, taken a few years later with a Polaroid filter that was rotated between successive photographs, show this effect, a striking and characteristic feature of synchrotron radiation.

The discovery of the synchrotron nature of the Crab's visible light introduced a problem that had no ready solution. Electrons with the needed energies would lose that energy so rapidly that after less than 100 years they no longer could radiate at visible wavelengths. Yet it was obvious that even after 900 years, the Crab was still a potent synchrotron source. A continuous supply of high-energy electrons was required, but none could be identified.

By now the Crab has been observed at all wavelengths, from radio to gamma rays, and a detailed picture has been constructed. The synchrotron radiation emerges from a glowing, irregularly shaped region, about one tenth of a degree across, corresponding to 10^{19} cm, more than half a million times the

Earth-sun distance. The abundance of many elements has been spectroscopi-cally analyzed and the debris identified as containing products of explosive nucleosynthesis. At the center of the Crab is a pulsar. This remarkable object, identified only in 1968, is rotating 30 times each second, beaming out its sig-nals (in all wavelengths) like a lighthouse. Researchers think the rotational ki-netic energy of the pulsar replenishes the electrons.

After it exploded in 1054, the Crab was bright enough to be seen by day for 23 days and by night for nearly 2 years. At the maximum, its luminosity was about 10^{41} ergs/sec, about 100 million times that of the sun. Today, more than 900 years later, the energy being radiated by the Crab, from several different

Figure 9.1. The Crab nebula, photographed through a polarizing filter set at different angles for each shot. The changes seen through this set of photographs show that much of the light is synchrotron ra-diation, produced by very high energy electrons spiraling in a magnetic field. (Photograph courtesy of the California Institute of Technology.)

mechanisms and through all its different wavelengths, still amounts to 2×10^{38} ergs/sec—almost 100,000 times more than the sun radiates. The total energy radiated over the 900 years amounts to 10^{50} ergs. Today the remnants of the Crab are moving away from the explosion center with speeds of around 1,500 km/sec; the kinetic energy of the outward-bound debris amounts to around 10^{49} ergs.

Current observations of supernovas are of interest not only for their probable connections with cosmic rays, but also as ways of testing theories of stellar evolution with their internal nuclear processes. The star that exploded as SN 1987A was a blue supergiant and is calculated to have had a mass in the range of 16–20 times the mass of the sun. Its original luminosity was around 10^5 times that of the sun. During the explosion, this increased by a further factor of 10^4, producing 0.07 solar masses of ^{56}Co (half-life 78 days), which was then a major source of the heating of the remnant. Acceleration of electrons was demonstrated by radio observations of a synchrotron spectrum with a slope of -0.95, implying an electron population with an energy spectrum of $E^{-2.9}$, similar to that of other cosmic ray particles. (Theory shows that electrons with an energy spectrum of the form E^{-a} will produce synchrotron radiation with a spectrum where the intensity at any frequency is proportional to that frequency to the power $-(a - 1)/2$.) Yet very high energy gamma rays have not been detected, in contrast to the observation of air showers initiated by 100 TeV electrons which have been seen coming from SN 1006. Beyond the evidence of fast particles, 20 neutrinos were detected.

If cosmic rays are produced mainly by supernovas, as in the Crab, how frequently do such explosions occur? This is not easy to estimate. Our view of large regions of our galaxy is obscured by interstellar dust, and we cannot even see clearly to the galactic center. Radio waves help to map SNR once we know they are there, but discovery of new supernovas rests primarily on visual sightings, and the most recent event in our galaxy was the 1604 supernova observed by Kepler, Galileo, and many Chinese astronomers. From historical records and modern observations of other galaxies where supernovas have been seen, it is now estimated that one supernova explosion occurs every 20–25 years in our galaxy.

The picture is further complicated by the fact that there are two main categories of supernovas. Type I supernovas occur roughly every 36 years and have greater luminosities than those of Type II, and their brightness diminishes more rapidly. Type II supernovas occur about every 44 years and probably represent the final stages in the evolution of stars appreciably more massive than the sun. The Crab is the remnant of a supernova of Type II. SN 1987A is also classified as Type II and may have involved a star initially of about 20 solar masses in which there was a collapse of the stellar core, with the resulting increase in pressure generating a shock wave that blew off the overlying material. Less than half of such a star might remain after the explosion, in the form of a rapidly rotating neutron star (Figures 9.2 and 9.3). The possible existence of this strange type of object was first suggested on theoretical grounds by Lev Landau in Russia in the early 1930s and by J. Robert Oppenheimer in the United States a few years later. It was thought to be about 10 km across with a magnetic field about 10^{12} times stronger than that of the Earth. The combination of rapid rotation and a strong magnetic field provides the ingredients for accelerating particles to CR energies. When pulsars were discovered, in 1967, there was at first no idea of their true nature. Oppenheimer's calculations had dealt with the nature of the collapsed star and not primarily with their radiation. The idea of identifying pulsars with rotating neutron stars was put forward by Thomas Gold of Cornell in 1968 and is now generally accepted.

Type I supernovas are thought to occur in the late stages of evolution of a binary system, where two stars move in close orbits around each other; one of the stars is a white dwarf and its partner is a normal star with an extended atmosphere. Material is attracted gravitationally to the surface of the dwarf star, where it accumulates. There is a well-established theoretical limit for the mass of a white dwarf, the Chandrasekhar limit, at close to 1.4 solar masses. Accretion of matter onto the dwarf ultimately produces a thermonuclear runaway when this limit is exceeded. No central remnant has been found in Type I SNR, and identification rests on radio and visible detection of the expanding shell of debris. SN 1006, 1572 (Tycho), and 1604 (Brahe) are all classified as Type I.

The supernova model for the origin of cosmic rays becomes more attractive when we examine the energy budget. Within the solar system—the only region directly accessible to us—the measured flow of cosmic rays shows that, on average, each cubic centimeter has cosmic ray energy content of $1-2$ eV, and we presume that this is typical of the rest of the galaxy. With a galactic disk volume of 10^{67} cm^3, the total CR energy content is then about 10^{67} eV, or 1.6×10^{55} ergs. As we saw in Chapter 8, the ^{10}Be, ^{26}Al, and ^{36}Cl isotope abundances indicate an average CR lifetime of 15 million years (5.4×10^{14} sec). Thus the rate at which CR energy is lost amounts to $1.6 \times 10^{55}/5.4 \times 10^{14}$, or 3×10^{40}ergs/sec. The energy requirements change if some of the galactic halo

Figure 9.2. Large Magellanic Cloud (LMC), shown before and after the outburst of Supernova 1987A. Left: 19 November 1960. Right: 24 February 1987, within a day after the explosion had been discovered. The brightness of this supernova was so great that a shorter exposure was needed, as can be seen by the faintness of the other images in this view. (Photographs courtesy of the South African Astronomical Observatory.)

volume is included, but this is still a good first estimate. If the galactic CR energy is to be maintained at a steady value, this loss must be compensated by the injection and acceleration of new cosmic rays.

If we take a conservative rate of one supernova in 50 years (1.5×10^9 sec) and assume that the typical supernova yields 10^{50} ergs in fast particles, then the average power, the rate of providing energy, is $10^{50}/1.5 \times 10^9$, or about 6×10^{40} ergs/sec. Using a supernova frequency of one in 25 years, we pick up a factor of two for the available energy. Although this calculation is admittedly very rough, it does provide a plausible agreement in terms of the availability of energy from supernovas: these enormous amounts of energy are easily adequate to power cosmic rays.

Another factor needs to be mentioned, and it will raise the source energy requirements. CR leakage out of the galaxy depends on the energy of the particles, those with the highest energy finding it easier to escape. As a result, the observed energy spectrum is steeper than the source spectrum, changing its exponent by about 0.6 to the observed $E^{-2.7}$. A flatter spectrum requires more energy input. In addition, the energy-dependent leakage introduces an energy

Figure 9.3. Hubble telescope photograph of the remnants of SN 1987A 11 years after the explosion. A shell of ejecta, moving out at high speed, is also colliding with previously expelled material, indicated by the arrow. (NASA Astronomy Picture of the Day for 17 February 1998; courtesy of P. Garnavich, Harvard Center for Astrophysics.)

dependence of the proportion of secondary CR particles produced by fragmentation en route, and this is indeed observed. This "leaky-box" model has been developed in great detail and remains our best synthesis.

Identifying supernovas as likely CR sources and demonstrating the adequacy of the energy they provide still leaves a major puzzle to be solved: How, in detail, is such a large fraction of this energy directed into CR? The enormous amount of energy released in supernovas comes mainly from gravity, during the process of contraction from a star with a typical radius of 10^{11} cm or more into a tiny remnant neutron star core having a radius of around 10^6 cm (about 6 miles).

Though the importance of supernovas for CR is not contested, the precise mechanisms and locations for CR acceleration are still unclear. In the Crab nebula with its pulsar, diversion of energy to CR takes place with an efficiency of more than 1 percent. In contrast, the highest efficiency achieved with accelerators such as those at Fermilab (near Chicago) and CERN (near Geneva) is a hundred times less. We detect e-m signals at many wavelengths, telling us that fast particles are streaming around the galaxy. Observation of nonthermal radio from the Crab and TeV gamma rays from SN 1006 indicates continued production of fast particles on a scale that is prompt by many cosmic standards. However, recent identification of ^{59}Ni nuclei among the heavy CR suggests that the picture is more complex.

The CRIS isotope spectrometer aboard the *Advanced Composition Explorer (ACE)* has been identifying individual isotopes as heavy as cobalt $(Z = 27)$ and nickel $(Z = 28)$, in the energy range 50–500 MeV/nucleon. (See Figure 9.4.) The importance of these nuclei is that they are heavier than iron $(Z = 26)$ and much less abundant in nature. When CR iron nuclei have collisions, the emerging fragments must be lighter than the original iron. Consequently, when we detect CR nuclei lighter than iron, they will include fragments from iron collisions as well as genuine primary particles that have arrived without suffering any collisions. In contrast, CR nuclei heavier than iron cannot have been contaminated because there are diminishing numbers of yet heavier nuclei whose fragments they might otherwise represent.

Most radioactive nuclei decay through internal changes and then emit an alpha particle, an electron or a gamma ray. In marked contrast, [59]Ni decays (with a half-life of 76,000 years) by capturing an electron and converting to [59]Co, but this can occur only if the [59]Ni has one or more electrons, as do the atoms we normally encounter. Acceleration of any atom to CR energies inevitably strips all electrons from the nucleus; thus the acceleration of a helium atom removes the two electrons, and we later detect the resulting nucleus, or alpha particle.

As has been shown in Figure 5.4, the stable isotopes of cosmic ray Fe (mass numbers 54, 56, 57, 58) are all cleanly identified. The cobalt isotopes 57 and 59 and nickel isotopes 58 and 60 (with traces of 62) are also resolved. The total absence of [59]Ni and the abundant presence of [59]Co tell us that any [59]Ni origi-

Figure 9.4. The isotopic spectrum of cobalt and nickel, as measured by the CRIS instrument carried by the *Advanced Composition Explorer*. (Courtesy of W. Robert Binns, Washington University, St. Louis, and *Astrophysical Journal Letters* 523 [1999]: L.61.)

nally present (as from nuclosynthesis in a supernova) must have had such low speeds as to retain or capture electrons, then change to become ^{59}Co. Had the acceleration of ^{59}Ni nuclei been prompt, they would have traveled as nuclei unable to decay and we should be detecting them. From the absence of ^{59}Ni, we infer, therefore, that there must have been a delay between injection and acceleration of at least 76,000 years. During this time, the Ni would have captured the needed electrons and converted to the Co that we detect.

This is an unexpected and remarkable discovery, and the delay contrasts with, for example, the situation in the Crab nebula and SN 1006, where acceleration has been taking place within the first thousand years after the explosions.

Turning now to the mechanism and location of acceleration, we can note that, as seen in the laboratory, combinations of high voltages and strong magnetic fields are needed to accelerate charged particles. The "handle" through which a cosmic ray can be accelerated is its electric charge. We know of no way to accelerate an uncharged (neutral) particle to CR energies. Where we detect neutral particles—neutrons, neutral pions—they must have appeared as a result of processes that first produced faster and charged parents.

We are familiar with electron acceleration in our daily lives, though this is not widely recognized. Most television sets use beams of electrons to trace out the patterns we see. Consider an even more directly accessible example: If you comb your hair rapidly in a dark room and then bring the comb close to a metal door handle or light switch plate, you can see a small spark. The combing has rubbed electrons off the comb or hair and so generated a voltage difference that can drive electrons through the air to create the spark. If you comb in the dark and watch in a mirror, you may see small sparks. Much larger voltages—tens of thousands times larger—are used in a cyclotron to accelerate particles for nuclear physics experiments. If a strong magnetic field is used to confine the particles to circular paths, the particles may make a thousand traversals of a 10,000-volt acceleration to reach an energy of 10 million (electron) volts, or 10 MeV. To attain much higher energies, such as those achieved at the largest particle accelerator—the Tevatron at Fermilab—protons are accelerated through a series of accelerating stages, each in its own vacuum. The final storage and accelerating ring has a diameter of almost 2 miles, with specially designed magnets to keep the circulating protons focused in a small beam. A final energy is 1 TeV, or 10^{12} eV.

A different accelerating mechanism is termed magnetic reconnection. If two plasma regions (ionized gas) collide and have oppositely directed magnetic fields, the collision region assumes a new configuration with lower energy, and with very strong electric fields. Particle acceleration can occur and absorb some of the energy that is being released. This mechanism is thought to be active in solar flares, but the theorists tell us that detailed, quantitative computations are extremely difficult.

Acceleration of CR occurs in several identified cosmic regions, for example in the Crab nebula and around pulsars, in sunspots, and in the solar wind. Supernovas represent one type of source in which rapid acceleration occurs in a relatively compact region. The ingredients are there, one might say, but the recipe evades us. On a larger scale, CR can be accelerated by moving magnetic fields and magnetized plasmas. In this process, first suggested in 1949 by Enrico Fermi, the random motion of CR relative to a moving magnetic field results in the particles' sometimes gaining energy and sometimes losing it. On balance the gains exceed the losses, and statistically a gain occurs with a slow overall acceleration. With the Fermi mechanism, energy gains are proportional to a particle's momentary energy, but radiation losses increase with the square of that energy. This sets a limit to the maximum energy that can be attained, which is not a serious problem for protons and heavy particles but is a problem in the case of electrons, which radiate strongly because of their small mass. High-energy electrons can, however, be produced through meson decay because the mesons are created with high energies and so are their electron decay products—the electrons do not themselves require acceleration.

Supernovas can also contribute to acceleration over an extended region, as shock waves spread out from their explosions into the interstellar medium. Shock acceleration of particles has been observed, but on a much smaller scale, in solar flares and in the interaction of the solar wind with the magnetic fields surrounding the Earth, Jupiter, and Saturn.

Energy spectra of CR near the Earth have been shaped by a combination of injection, acceleration, and propagation; these processes are modified by energy losses, collisions, and perhaps leakage from the galaxy, and they all depend on particle energies. Observations of solar CR from space probes and theoretical studies of acceleration mechanisms show that different acceleration processes can all produce CR with energy spectra having the characteristic E^{-a} shape, but with values of a that vary between models.

The combination of the chemical composition and the relative constancy of the energy spectra up to about 10^{14} eV suggests either that CR production is dominated by one type of process located within our galaxy, or that a blend of processes has smoothed away different histories. For particles above that energy, and certainly above about 10^{18} eV, the evidence favors assigning an increasing fraction of the cosmic rays to sources in external galaxies (but see Chapter 7). Intergalactic CR exchange is a natural consequence of the inability of the weak galactic magnetic field to contain ultra-high-energy particles.

Another puzzle in the search for acceleration mechanisms is how the highest-energy particles, those above 10^{18} eV, are produced. I noted in Chapter 6 the evidence for a change in the shape of the energy spectrum at that high energy, and this can be linked to another observation not previously introduced. Once we allow for the effects of the Earth's magnetic field, most cosmic rays are found to arrive fairly uniformly from all directions. For particles with high energies, this feature is studied by observing the arrival directions of air shower primaries. At energies up to 10^{15} eV, the variation in the number of particles arriving from different directions is less than one tenth of 1 percent. Departures from isotropy in arrival directions of high-energy showers have been reported, but different groups have disagreed about the preferred arrival directions. The AGASA group has reported observing an excess of CR from the general direction of the galactic center, but since this has not been independently confirmed, a note of caution is still warranted. It is notoriously difficult to identify systematic biases, but the importance of the basic question ensures continued interest.

At one time, as exemplified by the proposal of Baade and Zwicky, it was suggested that most cosmic rays came from outside the galaxy, but this idea was abandoned when the CR energy density was better evaluated and distances to other galaxies were more reliably determined. Potential CR sources are adequate to provide the energy to fill our galaxy with cosmic rays, but the energy problem would become unmanageable if the much greater volume between galaxies also had to be filled by CR to the same extent that we detect near Earth. An escape from this difficulty is provided if the only extragalactic CR are the very few with the highest energies.

There are additional reasons for no longer considering an extragalactic origin for most of the cosmic rays. First, the fragmentation of heavy particles and the resulting elemental and isotopic abundances would be different from those that we observe. Second, with the huge distances that would have to be traversed, the energy of the electrons would be degraded by collisions with many low-energy photons left over from the big bang. Most of these photons have wavelengths in the radio and infrared regions of the spectrum, and in fact the 1965 discovery of this radiation provided the single most important pillar for the big bang theory at that time. In the early stages of the big bang, the temperature was about 10^{12} degrees. Now, about 15–20 billion years later, the universe has expanded and the radiation has cooled to barely 3 degrees above absolute zero. This is also cosmic radiation (though it differs from what we have been considering), and its photons are everywhere, in and out of galaxies. Within galaxies the distances are small enough that photon collisions do not usually concern us, but they cannot be ignored when we start to consider CR traveling over intergalactic distances.

A third argument against the extragalactic origin of most CR is provided by high-energy gamma rays, described in more detail in Chapter 10. A major source of gamma ray photons above 70 MeV is the decay of neutral pions, produced in the collisions of protons above 700 MeV with the ambient interstellar hydrogen. Satellite mapping has shown a strong concentration of these gamma rays from around the galactic equator, whereas a more widely spread distribution would be expected if the primary protons were largely extragalactic.

Our search for an origin of the cosmic rays started with the set of abundances and the energy spectrum, from which we deduced the distance traveled, the age of the radiation, and the overall energy requirements. Although supernovas in our galaxy appear to serve as a good first-order model, there are still several loose ends. In seeking to identify CR sources, we must compare the abundances of the different nuclei in those regions not to the abundances of the CR observed near Earth, but to what we calculate the original abundances were, having corrected for fragmentation and loss en route. We then find that the observed abundances present a problem when they are examined in detail.

The steady improvements in particle identification and accumulated statistics, as achieved most recently by the detectors aboard the *Voyager, Ulysses,* and *ACE* spacecraft, for example, have shown that the charge spectrum for particles with charges 26 and above is not consistent with what would be expected on the basis of *r*-process nucleosynthesis alone. Instead there is a generally good agreement with the relative abundances found in the solar system, which represent a mixture of *r* process and *s*-process contributions. In the *s* process, nuclei only up to lead ($Z = 78$) can be produced, but in the more violent *r*-process the synthesis proceeds as far as uranium ($Z = 92$). As mentioned in Chapter 5, the observational evidence from space-borne detectors shows far more *s* than *r* nuclei. The solar system almost certainly formed when a dense region of the interstellar medium (ISM) contracted, and we do not know of any compact sources that are injecting a spectrum of solar-system-type particles. It seems, therefore, that we must search for an acceleration mechanism that works on general interstellar material.

In a qualitative sense, we expect acceleration to occur where supernova shock waves spread out and encounter atoms in the ISM, but quantitatively this model has a long way to go. A potentially supporting piece of information was recognized in about 1980. The interstellar material consists mostly of atoms, with some molecules and with only a modest degree of ionization, but cosmic rays are fully ionized nuclei without any attached electrons. It is widely assumed that the electrons are stripped from atoms during acceleration, for neutral atoms cannot be accelerated to CR energies. The energy needed to remove an electron from an atom is a few electron volts; the precise amount depends on the type of atom and how many electrons have already been removed. The energy required to remove the first electron is known as the first ionization potential (FIP), and there is a somewhat better correlation between CR abundances and FIP than with general solar system abundances; that is, where the FIP is high, there are fewer cosmic rays, and vice versa. This seems to indicate that atoms with lower FIP are preferentially accelerated because it is easier for them to become ionized and thus to gain the electric charge by which they can be linked to the magnetic field in an expanding shock.

Another anomaly (in a picture with many anomalies) is provided by some of the data on isotope abundances. Cleaner identification of isotopes of neon (Ne), magnesium (Mg), and silicon (Si), provided most recently by University of Chicago detectors on the *Ulysses* spacecraft, has shown excess abundances (at 60–80 percent levels) for the isotopes ^{25}Mg, ^{26}Mg, ^{29}Si, and ^{30}Si when compared to regular solar system abundances of the dominant ^{24}Mg and ^{28}Si. Stan Woosley at the University of California at Santa Cruz and T. A. Weaver at Lawrence Livermore Laboratory have pointed out that these data could be reconciled if the CR sources resided in stars whose initial C, N, and O abundances were 80 percent greater than the abundances in the solar system material; in that case the subsequent nucleosynthesis would produce the observed abundances. This interesting (but ad hoc) assumption unfortunately cannot yet explain the observation of the ^{22}Ne isotope among cosmic rays, where it is four times as abundant as in solar system material. What is shown by this sort of observation and calculation is that, with ingenuity, many of the initially discrepant numbers can be "explained," but one is often left with significant puzzles. Improvements in nucleosynthetic theory may yet resolve the isotope problem, or it may be that the accepted baseline of solar system abundances is not the best guide for clues to sources.

Are other CR sources possible? There are 100 billion stars in the galaxy, of many different types, and several of these types are possible candidates. Serious difficulties often quickly emerge, however. The large number of solar-type stars, for example, will surely have solar-type flares, but even during their most intense phases these cannot provide enough particles with high enough energies to account for CR. Rare stars, such as the Wolf-Rayet type, seem attractive as possible sources because they are extremely luminous and are losing significant amounts of mass, but we do not know whether they can or do produce CR. Ordinary novas probably produce some particles, but the numbers and energetics argue against their playing a major role. Another possibility is the galactic center, the scene of enormous activity, as revealed by infrared, X-ray, and gamma ray telescopes, but this is too far from the solar system to contribute much to CR observed near the Earth. And, of course, there are black holes, theoretically predicted as final stages in the evolution of massive stars and persuasively identified in at least the remnant of SN 1987A. These should be capable of providing sufficient energy, but the connection with cosmic rays has yet to be made.

Another sort of violent process is now the focus of increasing attention. In 1961, in U.S. satellites designed to look for clandestine nuclear weapons testing by the Soviet Union, bursts of gamma rays were detected. Kept secret by the military for many years, gamma ray bursts (GRB) were first systematically hunted by the BATSE system (Burst and Transient Source Experiment) on the *Compton Gamma Ray Observatory,* launched in 1991. GRB are an enigma—they are very short-lived, and seem to be at great distances and to involve great amounts of energy, but no general pattern of behavior has been seen. We would expect cosmic rays somehow to be associated but connections have yet to be made.

Thus, by exhaustion of alternatives, we keep returning to the supernovas. Although there are still numerous loose ends remaining, we have made progress. From the charge spectrum and the necessary fragmentation, the average 5 g/cm^2 path seems well established, though there are some observations suggesting a slightly longer path. The relative abundance of the ^{10}Be isotope has been used as a clock, and in this way, and with the observed energy spectrum, we have been able to determine an energy budget. The energy requirement rests on the reasonable assumption of galactic confinement (for most cosmic rays). Supernovas can supply the energy and some of the particles. Interstellar material is probably accelerated to mingle with those particles produced in compact regions. It has also been suggested that the expanding shock waves can accelerate particles from the surfaces of interstellar dust grains. The very highest energy particles (probably protons) come from outside the galaxy, but the acceleration process and sites for these have not been identified.

The combination of these different pieces of information has led us to the conclusion that most of the cosmic rays must be moving within the galaxy, and that only some of the relatively few with the highest energies can be extragalactic. Many of the remaining questions will be addressed by the AC-CESS system (Advanced Cosmic Ray Experiment for the Space Station), now being designed for the Space Station.

But we have not yet made use of the information on electrons, which, as we will see in the following chapter, supports the picture we have been developing.

10

COSMIC ELECTRONS AND GAMMA RAYS

So far I have focused almost exclusively on nuclear cosmic rays—the protons, helium nuclei, and heavier particles. Protons carry most of the CR energy, but heavy particles give information on composition, sources, and propagation. Electrons, as we will now see, can be exploited to yield complementary information. We are initially confronted with the same questions that arose with nuclei: Where do the particles originate, and what happens to them as they roam around the galaxy? The fate of the electrons hinges on major differences between their behavior and that of nuclei. Electrons do not combine to form heavier electrons, nor can they be fragmented. Because of their small mass, electrons are far more susceptible to losing energy in ways that produce radiation detectable on Earth. I will begin by reviewing the most important ways in which CR electrons can be produced and can lose energy.

Electrons are injected along with protons and heavier nuclei in outbursts such as supernova explosions. Shock wave acceleration of interstellar particles acts on electrons as well as on nuclei. Electrons picked up by these processes will all be negatively charged—the same type of electrons as in all normal atoms. To these are added electrons generated in interstellar collisions of nuclei during their propagation. The most plentiful of these collisions involve CR protons with interstellar hydrogen nuclei, usually referred to as proton-proton, or p-p, collisions. (The contribution from heavier CR primaries is similar to that of the protons, but rather smaller.) In these collisions, much of the kinetic energy of the projectile proton can be converted into roughly equal numbers of positively and negatively charged pions, along with much smaller numbers of more exotic particles. In about a fiftieth of a microsecond the pions decay into muons, which, after about two microseconds, decay into electrons and positrons. Meson production in this type of collision has been thoroughly studied at many accelerators, and the results reveal the energy spectrum of mesons and how their numbers increase with collision energy.

As noted in Chapter 5, cosmic ray "electrons" detected near the Earth include both primary and secondary electrons (e^-) and secondary positrons (e^+). In detecting "electrons," we do not always distinguish between the electrons and the positrons. Identification of "electrons" (+ and −) is often made by their common property, the production of an "electron-photon cascade" (Figure 10.1). The sign of the charge can be determined only in devices that use magnetic fields that separate the electrons and positrons by the directions in which their paths curve. It is now known that 10 to 20 percent of the primary "electrons" are positrons; the proportion depends on the energy.

As with the nuclei, the energy spectrum of the electrons (shown in Figure 10.2) can be represented by E^{-a}, and the changes of a with energy are clues to the history of the electrons. Electrons are far more sensitive than nuclei to the modulating effect of the solar wind. At energies below about 10 GeV, few traces are left near the Earth of the true interstellar electron spectrum, and it has been estimated (with large uncertainties) that the interstellar intensity of 100 MeV electrons is perhaps a hundred times higher than that seen near the Earth. At energies from 10 GeV and up, the modulating effect is not more than 10 percent, and interstellar inferences can be more safely drawn. At 10 GeV the energy spectrum has a slope of about $a = 3$, steepening to a value of 3.3 at higher energies. The overall number of electrons near the Earth is a few percent of the protons.

It is the processes in which electrons lose their kinetic energy that differentiate them sharply from the nuclei, and the basic factor is their small mass. Given equal forces, an electron will accelerate nearly 2,000 times as much as a proton. The acceleration or deceleration of a charged particle is always ac-

Figure 10.1. Photographs of successive stages in the development of a shower with an energy of 10^{15} eV, in which photons produce electrons, which in turn produce more photons until the energy of each particle is too low for further multiplication. (Photographs courtesy of Peter Fowler, University of Bristol.)

companied by radiation emission. For example, the sudden deceleration of electrons striking a metal target produces the X-rays used in medical radiography. For an electron moving along a curved path in a magnetic field, the radiated energy comes from the particle's kinetic energy, and synchrotron radiation is emitted (Figure 10.3).

For electron energies around 100 MeV, in a typical galactic magnetic field of 3 microgauss, synchrotron radiation is the dominant means of losing energy. The radiation emitted by each electron is spread over a range of frequencies but tends to peak at a frequency that depends on the strength of the magnetic field and on the square of the energy. Under these conditions, electrons of 100 MeV will radiate strongly at 0.48 MHz, those of 1 GeV at 48 MHz. Radiation at the lower frequency is difficult to detect because of the Earth's ionosphere, but 48 MHz is well within the range of radio astronomy. When syn-

Figure 10.2. Electron energy spectrum with experimental data points. At energies below 10 GeV, the intensity and shape of the spectrum are strongly influenced by the sun. The area attributed to Webber et al.is calculated from the synchrotron spectrum observed by radio astronomers. The solid line is calculated for a diffusion model assuming one supernova every 30 years. (Courtesy of S. Torii, Kanagawa University.)

chrotron radiation is produced by many electrons whose energy spectrum follows the familiar E^{-a} form, then the resulting radio spectrum will have the characteristic shape in which the intensity depends on frequency to the power $(a-1)/2$.

Cosmic radio waves were first detected accidentally by Karl Jansky in 1931, while he was studying radio static generated in thunderstorms (Figure 10.4). Jansky found that his newly discovered "signals" were not produced on Earth but instead seemed to be coming from the direction of the center of the galaxy. Radio astronomy mushroomed after World War II with major improvements in electronics, and it quickly became apparent that the cosmic radio "noise" (as it came to be known) had a spectrum totally different from the thermal spectra expected from hot astronomical objects. This "nonthermal spectrum" was later identified with synchrotron radiation.

By using synchrotron theory and mapping the cosmic radio spectrum, we can, in principle, deduce the energy spectrum of electrons whose radiation we are detecting. By aiming our radio telescopes in different directions, we can, in effect, look at electrons in almost all parts of the galaxy. The actual analysis is somewhat complex. Electron acceleration can also be caused by an electric field, such as an electron experiences briefly when it passes close to an atomic nucleus. The resulting radiation is known as brems (braking)-strahlung (radiation). Bremsstrahlung from a multi-MeV electron can produce X-ray and gamma ray photons. In interstellar space, CR electrons will typically travel a

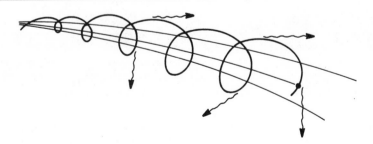

Figure 10.3. Electrons travel in spiral paths around magnetic field lines, continuously radiating photons whose energies depend on the electron energy and the magnetic field strength.

distance about the size of the solar system before having a sufficiently close encounter with an atom to produce bremsstrahlung. The result can be catastrophic for the electron, which can lose almost all of its energy in a single collision. In contrast, synchrotron radiation losses extends over galactic dimensions, and the electron radiates continuously. Each of the emitted synchrotron photons has low energy, and the electron's kinetic energy slowly

Figure 10.4. Karl Jansky and the radio telescope that he designed in 1929. This antenna was 100 feet in diameter and was tuned to a wavelength of 14.6 m (20.5 MHz frequency). Located at the Bell Telephone Laboratories in Holmdel, New Jersey, the whole system was mounted on four Ford Model T wheels, and rotated once every 20 minutes. Jansky was the first to detect radio signals coming from outside the earth. (Photograph courtesy of Bell Telephone Laboratories.)

trickles away. Bremsstrahlung is the most likely origin of low-level background gamma ray photons, where numbers have been mapped around our galaxy by several satellites, most recently NASA's *Compton Gamma Ray Observatory (CGRO),* launched in 1991.

A third way in which electrons can lose energy is known as inverse Compton scattering, in which a high-energy electron collides with (is scattered off) a photon (see Chapter 7). It is most significant for the highest-energy electrons, those above about 1,000 GeV (10^{12} eV). Inverse Compton scat-

Figure 10.5. Modern radio antenna; 27 of these dishes can be moved on rails that extend along the three arms of the Very Large Array (VLA), located near Socorro, New Mexico. Each of the VLA arms is about 20 km long, and each dish has a diameter of 25 m. The spacing of the dishes can be changed, depending on the measurements that are to be made. (Photograph courtesy of NRAO/AUI.)

tering has turned out to be important in several astrophysical settings. The many photons in interstellar space have come from two major sources. On average, there is about one starlight photon with an energy of a few electron volts in every 10 cm³. The energy density of these photons amounts to about 0.6 eV/cm³, quite close to the energy densities of cosmic rays and the galactic magnetic field. Many more photons are relics of the big bang cosmological explosion. Their energy density is about 0.3 eV/cm³, but since they have low average energy (about 0.0013 eV), there are around 200 of them in every cubic centimeter. These low-energy photons are found everywhere—all over the galaxy and also between galaxies. (Saying that there are 200 photons in a cubic centimeter does not mean that they are just sitting there. They are moving with the speed of light, but at any instant, 200 of them happen to be in any cubic centimeter.)

Inverse Compton collisions between electrons and cosmological photons can drastically reduce the electron's energy; the recoiling photon absorbs energy to reappear as a gamma ray photon. Calculations show that this type of collision effectively limits the observable electron energy spectrum to around 3,000 GeV. At higher energies, electrons are degraded so rapidly that few will ever be seen at the Earth. Indeed, the highest-energy electrons observed are in the region of 1,000–2,000 GeV.

To construct a general model for the CR electrons, we rely on direct observations of electrons and positrons near the Earth and add our detailed knowledge of the production of mesons in p-p collisions. From radio astronomical surveys, we can deduce the energy spectrum of distant electrons. We can allow for other ways in which electrons lose energy and adapt the leaky-box confinement model created for nuclear particles. What then emerges is a plausible picture. At injection, electrons have an energy spectrum proportional to $E^{-2.3}$. Because energy losses vary with electron energy, the emerging synchrotron spectrum will not have a simple form, but there is satisfactory agreement with what the radio astronomers observe.

Primary protons with energies above 10^{15} eV can produce mesons, which in turn produce electrons with energies well above 1,000 GeV, but there will be few of these because so few protons have the necessary energy. Even with such high electron energies at production, inverse Compton losses produce an effective cutoff to the electron spectrum. Synchrotron losses dominate down to energies of 10 GeV; at lower energies escape from the galaxy becomes important. For electrons with an energy below about 1 GeV, the solar wind modulation is so large that we lack reliable information on the interstellar electron intensity, which may be many times larger than we observe near the Earth.

A good fit between the synchrotron radiation spectrum and the near-Earth electron observations is obtained by using an average interstellar magnetic field of 5 microgauss and a lifetime that varies between 30 million years at 1 GeV and 4.8 million years at 100 GeV. These cover the range of lifetimes deduced from the ^{10}Be and heavier isotope data. The observed electron/positron ratio is consistent with production in p-p collisions, provided the proton path length is about 10 g/cm^2, which is somewhat larger than that deduced from the production of L nuclei by C, N, and O nuclei.

Further observations and modeling are clearly needed to reconcile the differences between quantities such as field strength or path length as these are deduced from different observations. The injection processes still remain a large area of uncertainty. We know that electrons are accelerated in the Crab nebula and other supernova remnants, with and without pulsars, and, on a much larger and more distant scale, we detect synchrotron radiation emerging from quasars (probably cores of active galaxies). The web of connections with the conventional cosmic rays remains to be firmly established.

Let us now turn from the low-mass electrons to the photons, which do not have any mass in the conventional sense. They do not respond to Newton's formula $F = ma$, for no force can accelerate a photon; its speed is always that of light. Still, photons carry energy and momentum, and these can change when photons collide with particles.

The distinction between particles, photons, and waves has been blurred for many years. This issue has provoked controversy since at least the late seventeenth century, when Newton and Christiaan Huygens championed competing particle and wave theories to explain the nature of light. Quantum mechanics and the less well known quantum electrodynamics constitute a comprehensive modern theoretical framework that has been extraordinarily successful in quantitative application, although researchers often (for convenience) use the older and simpler particle or wave models for specific problems.

Where, then, should we draw the line in defining cosmic rays? By now they are known to be particles, although a photon model was initially strongly advocated by Millikan. If we were to include all photons in the definition, we would have to include all of astronomy, which is not practicable here. Accordingly, we extend our boundaries to cover only the high-energy gamma ray photons produced by and associated with CR particles. We can use those photons as diagnostics of distant CR, as we have just done for electrons with the radio spectrum. Even here, we will need to be selective, for gamma ray astronomy is vigorous and more extensive than we have room to survey here.

Photons with very high energies can be traced to the creation of pions in interstellar p-p collisions. From the same collisions that produced charged pions, neutral pions also emerge and promptly decay into gamma rays detectable over a range of energies peaking broadly in the range 70–100 MeV. While production of neutral pions in the Earth's atmosphere quickly leads to electron-photon cascades, in the extremely low density of interstellar space no such showers can develop. Photons from neutral pions therefore travel across the galaxy with little chance of absorption.

Extensive celestial surveys of high-energy photons were carried out by two satellites, NASA's *SAS-2* and the European *COS-B,* launched in the 1970s. *SAS-2* detected 8,000 photons. *COS-B,* during its lifetime of nearly 7 years, detected 210,000 photons, about as many as the number of less energetic photons we receive by eye in 40 seconds from the faintest visible star. These surveys have been far surpassed by NASA's *Compton Gamma Ray Observatory (CGRO),* launched in 1991. With four instruments aboard, *CGRO* covered a wide range of photon energies. The highest-energy gamma rays examined by this mission have been recorded by the EGRET instrument (Energetic Gamma Ray Experiment Telescope), designed to detect photons in the range between 20 MeV and 30 GeV (Figure 10.6). Another instrument, BATSE, has the capability of responding to rapid changes in intensity. The subject of these gamma ray bursts is continually in the news, but, though fascinating, lies outside our coverage here.

In early 2000, failure of one of the stabilizing gyros forced NASA to review the *CGRO* mission. Operation of the EGRET detector required periodic gas refills and the last was being nursed for special viewing; the other three instruments were still active. NASA's concern was that since *CGRO* weighs 16 tons, if the satellite had to be brought down in a crash reentry into the atmosphere, full control was needed in order to avoid populated areas. At the time of this writing, the options being examined included bringing CGRO down soon and before another gyro failure occurred, or leaving it in orbit but re-

Figure 10.6. The EGRET detector for high-energy gamma rays, carried on the *Compton Gamma Ray Observatory.* Charged particles are rejected from further analysis by signals they produce in the anti-coincidence dome. Gamma rays convert into electron-positron pairs in the upper chamber and the gamma ray energy is calculated from the subsequent development of the shower in the lower chamber and the sodium iodide (NaI) counter below. (Courtesy of D. L. Bertsch, NASA Goddard Space Flight Center; *Astrophysical Journal* 86 Supplement [1993]: 629.)

trieving it via the Shuttle, which had placed it in orbit originally. Keeping *CGRO* in orbit permits the unique BATSE to continue monitoring gamma ray bursts, for there is no other instrument ready to take its place.

A celestial gamma ray map has been constructed from EGRET data (Figure 10.7), and it shows that the diffuse emission related to cosmic rays and their pions is clearly concentrated in the galactic disk. Several conspicuous compact regions of high intensity can be identified with known objects such as the Crab nebula and the strong X-ray source Cygnus X-3. The intensity of diffuse high-energy gamma rays agrees with that expected if the general galactic cosmic intensity is similar to what we detect near the Earth.

Researchers are now exploring a new region of gamma ray energies by examining showers (EAS) that are produced by photons rather than by nuclei, such as the protons and their heavier cousins that have been described in Chapters 6 and 7. Here we are concerned with primary photons with energies in the TeV range. A gamma ray with sufficiently high energy will convert into an electron-positron pair in the upper atmosphere, typically above 10 km altitude; each of these new particles will in turn emit high-energy photons (in the bremsstrahlung process), and these will themselves convert to electron-positron pairs; this process continues through many generations as the shower develops. As they speed downward, high-energy electrons will emit Čerenkov radiation and millions of their photons can be detected at ground level.

The Čerenkov radiation is in the blue end of the visible spectrum. It is collected by telescopes and brought to a focus on photomultipliers. As with optical astronomy, detection is strongly directional so that objects or regions known or suspected to be sites of energetic particles and processes can be examined and tracked as the Earth rotates. Also as with optical astronomy, very clear and cloud-free skies are needed, so that desert sites are preferred. For example, the 10m Whipple telescope of the Harvard-Smithonian Astrophysical Observatory is located on Mt. Hopkins in Arizona. (Fig. 10.8). Other atmospheric Čerenkov gamma ray observatories have been set up in both the southern and northern hemispheres. There is, however, an important way in which this new gamma ray astronomy differs from conventional optical astronomy. A typical shower of Čerenkov photons has spread out over an area

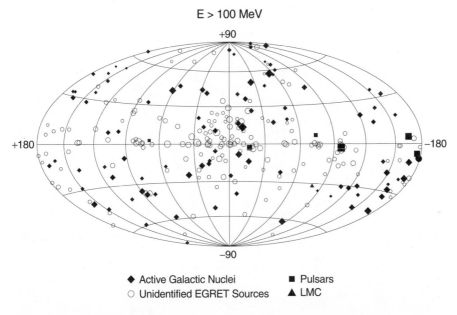

Figure 10.7. Survey of high-energy gamma rays as observed by the EGRET detector aboard the *Compton Gamma Ray Observatory.* Top: The Milky Way galaxy, showing contours of equal brightness in the direction toward the galactic center. Note the strong source of gamma rays near 280°, identified as the Vela pulsar. Bottom: Point sources of high-energy gamma rays, shown as open circles on this map of the sky. For comparison, the locations of objects already identified as sites of great activity are also shown. These include Active Galactic Nuclei (AGN), pulsars, and the Large Magellanic Cloud. It is clear that gamma rays cannot be identified with any of these classes of objects. In addition, many of the gamma rays come from directions well away from the galactic equatorial plane. (Courtesy of D. L. Bertsch, NASA Goddard Space Flight Center; *Astrophysical Journal* 123 Supplement [1999]: 79.)

of more than 10,000 square meters at ground level, and a central "hit" is not needed for shower detection. The arrival of a shower is sampled by a telescope whose effective "collecting area" is thus much larger than that of the mirror dimensions alone, and the energy of the original photon can be calculated from a reconstruction of the shower.

In this way, gamma radiation at TeV energies has been detected coming

Figure 10.8. The Whipple telescope, on Mount Hopkins in Arizona. Čerenkov light is collected by 109 mirrors in an array with a diameter of 10 m and brought to a focus on the same number of photomultipliers. (Photograph courtesy of Trevor Weekes, Harvard-Smithsonian Astrophysical Observatory.)

from the Crab nebula, from the active galaxies Markarian 421 and 501, and from a pulsar. Possible additional identifications have been reported but not yet confirmed. Astrophysical events and processes that escape observation by conventional methods might well be accessible through this approach.

Much lower energy gamma rays, in the range of a few MeV, can be produced by a different set of CR interactions. Protons with energies below 100 MeV cannot produce mesons, but when they collide with nuclei some of their kinetic energy is transferred and the nuclei become excited. When one of these nuclei returns to its normal state, the excess energy is emitted as a gamma ray with a sharply defined energy. (This differs from the situation with synchrotron radiation and bremsstrahlung, where photons can have any energy over a wide spectrum.) Proton collisions in dense regions around or between the stars are expected to yield gamma rays characteristic of certain nuclei and thus to be easily identifiable. In oxygen and carbon atoms that are expected to be plentiful in dust grains, proton bombardment can produce gamma rays of 6.1 MeV and 4.4 MeV, respectively. Photons with an energy of 1.809 MeV, attributed to the decay of ^{26}Al, have been detected, with a concentration from the direction of the galactic center. A particularly interesting gamma ray line has an energy of 0.511 MeV, from the annihilation of positrons and electrons, and this, too, has been detected from the galactic plane with a concentration from the galactic center direction. The overall picture is one of great activity throughout our galaxy, very different from what might be expected from a tranquil picture of stars twinkling in the night.

We may now sum up, recognizing that there are still many loose ends: CR electrons have been detected at Earth, and their energy spectrum can be understood through their association with the nuclear CR, though their low abundance compared to that of the protons remains a puzzle, considering that there are as many positively as negatively charged particles in our galaxy. The presence of energetic electrons in distant regions is well established through surveys of synchrotron radiation. Similarly, some gamma rays, which can only have come from CR interactions, have been observed. All of these provide information complementary to what we obtain directly from CR nuclei, and allow us to diagnose conditions far from Earth. We can now turn to the last in our list of the cosmic "particles" and the most elusive—the neutrinos.

11

NEUTRINOS, LIKE PHOTONS, carry no electric charge and were, until recently, thought to have zero mass. The first suggestion that neutrinos exist came in 1930 from Wolfgang Pauli, a theorist. Measurements of the energies of electrons that were ejected from some radioactive nuclei seemed to suggest that total energy and momentum were not held constant ("conserved") in those nuclear processes. Such conservation laws are central principles in physics. To preserve them, it was Pauli's ingenious but radical suggestion that some undetected particles were also being emitted along with the electrons—and these undetected particles had neither electric charge nor mass but still carried energy and momentum. Fermi dubbed them "neutrinos"—the little neutral ones.

Pauli was so renowned for his brilliance as a theorist that he was called "Zweistein," and he was greatly bothered by the idea of particles with such outlandish properties. He also suggested that atomic nuclei should contain neutral particles in addition to the protons that each carried a positive electric charge. Such "neutrons" could remain indefinitely within an atomic nucleus, but if they came out they would be radioactive and each would decay into a proton, an electron, and a neutrino. Pauli's work was recognized through the award of the Nobel Prize in 1945.

Since the early days, experiments and their theoretical interpretations have become much more sophisticated, and we now know that there are actually three different types of neutrino, each with its own antineutrino partner (much as electrons have positrons as their antiparticles). These three types are termed the electron neutrino, the muon neutrino, and the tau neutrino.

The galaxy is awash in neutrinos, which pass through us without any known biological effect. Some neutrinos come from the Sun, necessary by-products of the nuclear reactions at its core; other stars produce their share of neutrinos but because they are much farther away, we do not take note of them at Earth until an eruption occurs, as with a supernova. In addition, neutrinos are produced as the secondary products of cosmic ray collisions in the Earth's atmosphere. Cosmic ray energy is converted into pions that live for around one hundred millionth of a second before they decay to muons, each accompanied by a neutrino. The longer-lived muons (they live for two millionths of a second) decay into electrons plus two neutrinos.

From this variety of sources, neutrinos bombard the Earth at all times and evenly from all directions. Because of their relation to nuclear and particle processes, cosmic neutrinos can be regarded as a component of cosmic ray physics, and only the technical difficulties of neutrino detection have slowed the development of this field.

The sun's radiant energy comes from nuclear reactions in its core. In that central region, the enormous pressure produces a temperature of 15 million degrees. The positively charged hydrogen nuclei are then moving rapidly enough to overcome their electrostatic repulsion and they fuse to produce helium. Through a series of nuclear reactions, four hydrogen nuclei (protons) form a helium nucleus, and a small amount of their original mass is converted into the energy ($E = mc^2$) that keeps the sun aglow.

Although the starting point in the fusion reactions always involves hydrogen, there are several different sequences of nuclear reactions that lead to the production of helium. Along the way, nuclei of beryllium, ^{7}Be, and boron, ^{8}B, are produced and promptly consumed. Although these nuclei do not survive, the different reaction chains produce different types of neutrinos with different energies. Thus neutrinos from collisions between two protons with resulting fusion have energies below 0.5 MeV, those from the ^{8}B chain cover a range of energies extending up to just over 10 MeV, and neutrinos from the ^{7}Be chain have sharply defined energies of 0.3 and 1.0 MeV. Neutrinos produced in meson decay will have far higher energies, derived from the energies of their parent mesons.

The neutrinos produced in these different processes differ in how they themselves can subsequently interact with other particles. For our purposes, it will suffice to note that there are these three types but we do not need to venture into the arcana of their differences.

Neutrinos carry energy and momentum and can interact with electrons and nuclei, but seem to do so with the utmost reluctance. The likelihood of a particle's interacting is often expressed as a cross section, the effective target area it presents to other particles. For example, a cross section for one of us being hit by a tennis ball might be a few square feet, but for an atom a typical cross section is 10^{-15} cm^2. The typical unit for nuclear capture of slow neutrons, leading to fission, is around 10^{-24} cm^2, which Fermi characterized as being as big as a barndoor—hence the term millibarn (or mb) for 10^{-27} cm^2. For neutrinos of MeV energies, the cross sections run around 10^{-43} cm^2, or 10^{-16} mb, but the cross section is strongly energy dependent. In general, a solar neutrino has only a few chances in a trillion of having a collision even while passing diametrically through the Earth. The process of detecting neutrinos is difficult and inefficient. Nevertheless, they are important members of the family of elementary particles, and the elucidation of their properties has kept a lot of high-energy physicists very busy.

Although the observation of charged particles in various meson decays gave strong support for the existence of neutrinos, the first direct detection came in 1956. The discoverers, Clyde Cowan and Fred Reines, worked first in the state of Washington at the Hanford nuclear reactor, which produced copious numbers of neutrinos. With encouraging results, they moved their experiment to the more powerful Savannah River reactor in South Carolina. There they were able to count about 2.5 neutrinos per hour, sufficient for definitive results. (Because of the ephemeral nature of the neutrinos, they termed their quest Project Poltergeist.) Neutrinos produced under different laboratory conditions have now been studied for more than 40 years, and our knowledge of the properties of the neutrinos is based on the results of these experiments as well as on studies of neutrinos being detected from astronomical sources.

Striking but puzzling results have come from cosmic neutrino observations. For over 30 years, Raymond Davis of the Brookhaven National Laboratory has operated a solar neutrino detector in the Homestake Gold Mine in Lead, South Dakota, 1.5 km underground (Figure 11.1). Because of the expected low neutrino counting rate, this location was selected to minimize the ever-present CR background. The detector's core is a tank holding 400,000 liters (about 100,000 gallons) of extremely pure perchlorethylene, a cleaning fluid. Solar neutrinos react with some of the fluid's chlorine atoms to create ^{37}Ar, an isotope of argon, which is radioactive with a half-life of 35 days. To detect the neutrino-produced argon, Davis flushed the system with helium gas every 100 days to extract the argon. During those 100 days many of the argon atoms decayed, but the radioactivity of the survivors could still be measured.

The sun's neutrino production can be calculated from the theory of stellar structure and nuclear reactions, generally considered to be well established and described in the Standard Solar Model (SSM). Just over 10^{37} neutrinos are calculated to leave the sun every second, and at Earth the flux is about $10^{10}/$ cm^2.sec. But for every 10^{21} neutrinos that penetrate Davis' neutrino detector, only one will induce a chlorine atom to change to argon, and the expected rate of production was about three atoms per week. A solar neutrino unit (SNU) is defined as the number of captured neutrinos per 10^{36} target atoms, and in these units the predicted counting rate was 7.5 $\pm$ 3.3 SNUs. But Davis' observed rate, though it varied a bit over the years, had an average of 2.55 SNUs, about one third of the predicted rate.

The importance of this large discrepancy between observation and theory cannot be overemphasized. Modern stellar theory has been widely accepted, founded on principles that were thought to be well tested. Because this theory has brought order to an otherwise chaotic diversity of astronomical observations, there is great reluctance to abandon it. Yet there is clearly a problem when our closest and best studied star refuses to conform.

All sorts of explanations have been considered in an effort to solve the solar neutrino problem. The accepted solar model has been constructed to fit the observed luminosity and size of the sun, both beyond question. But the light energy coming from the solar surface originated more than a million years ago in the core's nuclear reactions. The neutrinos, by contrast, because of their

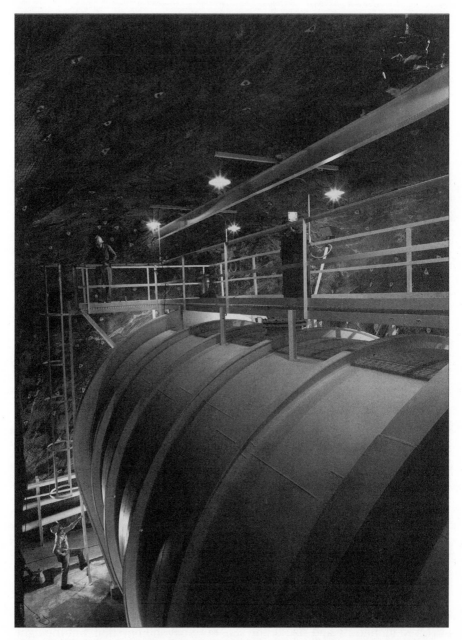

Figure 11.1. Tank containing 100,000 gallons of the neutrino detector perchlorethylene in the Homestake Gold Mine, South Dakota. (Photograph courtesy of Raymond Davis, Jr., and Brookhaven National Laboratory.)

obliviousness to other particles, travel directly from the sun's core to Earth in 8 minutes. Could the neutrinos, in telling us what was occurring only 8 minutes ago, be foretelling the surface behavior of the sun a million years in the future? Alternatively, could neutrinos be more complex particles than we had believed?

A possible answer to these questions brings in the issue of neutrino mass, long taken to be exactly zero, as was originally assumed and has since been supported by experiments and calculations. If, instead, neutrinos had finite, though very small, mass and came in more than one type, there could be significant consequences, such as oscillation of neutrinos between different types. Neutrino oscillation was suggested by S. P. Mikheyev and A. Yu Smirnov and independently by Lincoln Wolfenstein, in what is now termed the MSW model. Oscillation has the effect that neutrinos of one type emitted in a nuclear reaction could change to another type during travel to the Earth. Some neutrino detectors can detect all types of neutrino, but others respond only to one or another type. As a result, an apparent deficit in the number of neutrinos seen by one particular detector might be the indication of neutrino oscillation (into an undetectable type) and thus can be an indicator of neutrino mass.

Davis' pioneering work has been followed by several other neutrino detection programs. Because of the microscopic cross section for neutrino interactions, the sizes of these detectors are on a gigantic scale. Thus the Superkamiokande detector, located in the Kamioka mine 200 miles west of Tokyo and about 1000 m below the surface, contains 50,000 tons of ultra pure water (Figure 11.2). Neutrino detection in Superkamiokande occurs only after a neutrino collides with an electron in the water and ejects it from its parent molecule. For a neutrino of high energy, the knock-on electron travels fast enough to emit Čerenkov radiation in the form of visible light, which can be detected; ultrapure water is needed to minimize absorption of this Čerenkov light.

In Davis' experiment, the neutrino interactions leave their signal in the form of radioactive ^{37}Ar nuclei which are later extracted by a chemical process, but any information on the direction in which the neutrino had moved has been lost. In Superkamiokande, however, we make use of a special property of

Figure 11.2. Left and right: Superkamiokande neutrino detector, showing some of its 11,146 photo-multiplier tubes (PMTs) and a crew during installation before complete filling. (Photograph courtesy of the Institute for Cosmic Ray Research, University of Tokyo.)

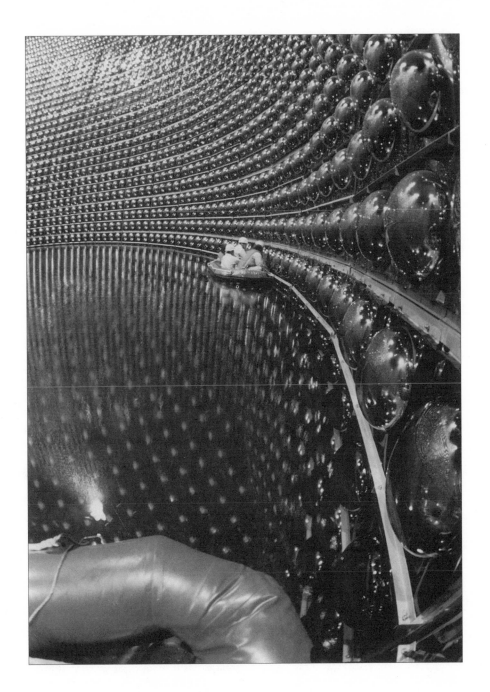

Čerenkov light: it is confined to a narrow cone around the direction of movement of the parent particle. The direction of the radiating electrons can thus be detected and that tells the probable directions of the original neutrinos.

The directionality of the Čerenkov light is then put to good use. In Superkamiokande, some of the system's 11,146 photomultiplier tubes (PMT) are mounted so as to detect light from downward-moving electrons. Other PMTs are mounted so as to detect light produced only by upward-moving electrons, which must have been produced by neutrinos that have traversed the entire diameter of the Earth before entering the water and knocking on an electron in the upward direction. Upward-moving neutrinos have had to travel farther and so have had time to undergo more neutrino oscillations; and the occurrence of oscillations will thus be indicated by a difference between the up and down counting rates.

In 535 days of operation, Superkamiokande collected over 4,000 neutrino events, with 236 definitely moving down and 139 moving up. The average electron energies were somewhat over 10^9 eV and the observed difference between upward and downward fluxes can best be explained through oscillation between mu and tau neutrino types, with the tau not being detectable. A different oscillation process, between mu and electron neutrinos, does not fit the data nearly as well. The neutrino mass calculated from the Superkamiokande data amounts to at least 0.1 eV, or one fifth of a millionth of an electron mass.

As this is being written, other neutrino detecting systems are in varying stages of planning, construction, and testing, so that we may hope for a clearer picture in a few years. Of necessity, these experiments are so large that they require collaboration between many laboratories or research institutes, often across national boundaries. Thus a U.S., British, and Canadian collaboration has constructed and now operates the Sudbury Neutrino Observatory (SNO) in Ontario, in a nickel mine, 2 km underground. Designed to detect cosmic neutrinos of all types but not their directions of motion, the sensitive volume houses 1 million kilograms of heavy water (D_2O), in which the oxygen atom is combined with two atoms of the heavy hydrogen isotope deuterium. Some collisions of solar neutrinos can fragment the deuterium nucleus (deuteron) into two protons and an electron. The electron is fast enough to lose energy by radiating Čerenkov light, which is detected by 10,000 photomultipliers that were installed by a remotely controlled submarine. In a later stage, the system will look for other neutrino collisions, in which the deuteron can be split into a proton and a neutron. To detect the neutrons, sensors will be added that contain ^{3}He, the light isotope of helium. Neutron capture by ^{3}He will produce the major isotope ^{4}He.

Perhaps the most exotic neutrino observatory is AMANDA (Antarctic Muon and Neutrino Detector Array) where light detectors have been lowered to a depth of 2 km in holes drilled in the Antarctic ice. As in Superkamiokande and SNO, neutrino detection in AMANDA depends on the observation of Čerenkov light from high-energy electrons knocked on by neutrinos. Again, the direction of the emitted light preserves the original neutrino direction to within about a degree. At great depths, the Antarctic ice is so clear that the Čerenkov light can travel considerable distances without absorption. The sensitive volume being monitored is cylindrical, half a kilometer high and 120 m in diameter. AMANDA is sensitive to neutrinos with energies higher than those detectable by any of the other systems, up to 10^{12} eV. Cosmic neutrinos will be detected, along with those that come from collisions of cosmic ray particles in the atmosphere that lead to neutrinos via the pion to muon path. At AMANDA's great depth, neutrinos and muons can be detected, but since AMANDA uses 300 phototubes that face downward, only the neutrinos that have traversed the Earth's diameter will be detected.

With all of this effort being expended to detect neutrinos, what are the questions that we can hope to answer? There are several. First is the nature of the neutrino itself—does it or does it not have any mass? If further observations contradict the Superkamiokande results and, with reduced experimental errors, can rule out the possibility of neutrino mass, then the theoretical models of stars may need serious revision. On the other hand, if Superkamiokande's results are confirmed, then neutrino oscillation must occur and the neutrino must have mass, and we can have added confidence in the stellar models. But once neutrino mass is definite, then the precise value of that mass is of great importance. Many other astronomical measurements have long suggested that there is far more matter (and mass) in the universe than we can see when we count up the stars, gas, and dust. What is the nature of this "missing mass"? Might the neutrinos, in their vast numbers, supply this missing mass? Also, as we know from the spectra of galaxies, our universe is expanding—the more distant a galaxy, the faster it is moving away from us. What is not known is whether the universe is open or closed, that is, whether the expansion will continue forever, or whether it will, sometime in the remote future, slow down, then stop and reverse itself to contract with ever-increasing speed, reversing the big bang to create the big crunch.

What will determine the fate of the universe is its total mass. With sufficient mass, gravitational attraction will ultimately slow the expansion to a halt and then force a reversal into a contraction. There are so many neutrinos in the universe that their contribution to the critical mass could be significant if they each had some mass, even a very small one. Present Superkamiokande data point to a neutrino mass less than 10^{-6} times the electron mass. The challenge is to see whether this experimental limit can be pushed down and, if so, how far.

In February 1987, when supernova 1987A appeared, scientists were able to detect neutrinos close to the time of the visual outburst. So reluctant were neutrinos to interact with other particles that of the more than 10^{16} that were calculated to have streamed through detectors on Earth, only 19 neutrinos were detected, fewer than the number of scientific papers since written about them. In one neutrino detector system, operated near Kamioka, Japan, by a Japan–University of Pennsylvania collaboration, 11 neutrinos were detected over an interval of a few seconds. The other 8 neutrinos were detected in a system near Cleveland, run by scientists from the University of California at Irvine, the University of Michigan, and Brookhaven National Laboratory.

The precise time of the onset of optical supernova flaring cannot be pinpointed to better than a few hours, so any delay of the neutrinos' arrival relative to the outburst is not well known. What can be pinpointed very precisely is the spread among the neutrinos arrival times. Particles with zero mass will all travel at exactly the speed of light. If the neutrinos were all emitted in less than a second (as is suggested by some theorists), they should all have arrived within a similarly small interval. If, however, the neutrinos have nonzero masses, then, like electrons or protons, their speeds will depend on their energies, and a spread in energies will show up as a spread in arrival times, as was seen. The 19 neutrinos were detected over an interval of 12 seconds, and from this and the measured neutrino energies, the neutrino mass could be calculated.

Different theorists, using different assumptions, have arrived at slightly different values for the neutrino mass, but none is more than 1/50,000 of an electron mass. This is a major discovery, but so small a neutrino mass is hardly sufficient to close the universe. The problem should not yet be considered solved; it is possible that present assumptions about the supernova explosion itself, as well as about the behavior of neutrinos within the exploding star, might change.

We are left with tantalizing appetizers, waiting for the next supernova explosion. It has often been remarked that astronomy is an observational but not an experimental science, and this is no better illustrated than by the neutrinos. The essential element of experimental science is missing: we cannot control the conditions of the "experiment." We are unable to adjust the sun's temperature and wait to see how this affects the flow of neutrinos; we cannot arrange for a supernova to explode in a star of known mass. Our ingenuity is confined to the design of our apparatus and the sophistication of later analysis. In the meantime we exercise great patience.

12

THE SUBNUCLEAR WORLD

Over the years, cosmic ray studies have seen several shifts of emphasis. Currently most attention is directed toward the astrophysical aspects, but for a period of about two decades, cosmic rays provided the only way of discovering new subnuclear particles. Today, the study of these particles is carried out almost exclusively in large international centers such as the CERN laboratory in Geneva and at Fermilab near Chicago, with their proton accelerators, and Stanford Linear Accelerator Center (SLAC) at Stanford and the Deutsches Electronen Synchrotron (DESY) in Hamburg, with their electron accelerators. Occasional CR observations may indicate the presence of new particles in an energy range still beyond any machine's ability, but this is no longer where the particle action is. It is interesting to note that the almost industrial scale of today's experiments in particle physics has been a factor behind the switch of a number of prominent particle experimenters to CR research.

When cosmic radiation was discovered, views of the structure of matter were simple. The catalog of the elements had been based on chemical separation of materials found in nature. Each atom of each element was supposed to consist of a central nucleus carrying a positive electric charge, surrounded by a number of negatively charged electrons. The atom as a whole was electrically neutral. Within the nucleus were thought to be protons carrying positive charges and a sufficient number of electrons to bind the protons together; otherwise the collection of like-charged protons would soon disperse. No other (nonelectric) forces were known or invoked.

Although it did not provide the answers to all questions, this atomic model survived the invention of quantum mechanics in the 1920s and lasted until Pauli's 1930 theory and James Chadwick's discovery of the neutron in 1932. Chadwick used alpha particles emitted from radioactive polonium to bombard a thin foil of beryllium. Neutrons ejected from the beryllium nuclei were identified as they collided with protons whose recoils could be detected. The typical alpha particles that Chadwick used, from radium or polonium, had energies of a few MeV, far below CR energies and thus very limited in the nuclear effects they could produce. Understandably, therefore, subsequent particle discoveries came first from the study of cosmic rays and later from experiments with more energetic particle accelerators.

Working at almost the same time as Chadwick, Carl Anderson and Seth Neddermeyer of Cal Tech were encouraged by Millikan to pursue their study of the penetrating power of certain CR particles. In 1932 they used a cloud chamber surrounded by a giant electromagnet to identify CR particles and measure their energies; the discovery of positrons rested on seeing their curved paths in the magnetic field. It was just at this time that Blackett and Occhialini found, with a counter-controlled cloud chamber, roughly equal numbers of positrons and electrons and thus provided the first confirmation of Dirac's theory of particles and antiparticles.

Thus by the end of 1932 the list of particles had grown. Nuclei were known to consist of protons and neutrons. Dirac's particle and antiparticle theory had passed a major test, and there was an increased theoretical attack on the problem of nuclear stability. The electric force between two particles depends inversely on the square of their distance apart (d). Within the tiny confines of a nucleus (with a diameter less than 10^{-12} cm), this $1/d^2$ repulsive force between protons should increase to overwhelming proportions. What sort of force holds the charged protons together against their mutual electrostatic repulsion and also binds them to the uncharged neutrons? Although there were several other major questions regarding the nucleus, the central one in the 1930s was still that of binding.

The key to the understanding of nuclear forces came from a new idea, that of exchange forces. An analogy used by the theorist Paul Matthews of Imperial College in London, is more helpful than a definition: Imagine two children playing with a ball, tossing it back and forth. The distance they can stand apart will clearly depend on the weight of the ball. If they play with a tennis ball, they can stand far apart, but playing with a basketball keeps them much closer together. We can say that they are bound together by the exchange of their ball. At times A will hold the ball, at times B has it, and at times it is in transit, being exchanged. The range of an exchange force varies inversely with the mass of the exchanged object. In theoretical models of the electromagnetic force, the photon, that bundle of energy, is the exchanged object, the "ball" that holds an electron to its positively charged nucleus or causes the deflection of one charged particle passing near another. Because the photon's weight is zero, the range of the electromagnetic force is great. The force, with its $1/d^2$ dependence, extends to infinite distances, although its strength does diminish to a vanishingly small value.

Hideki Yukawa at Osaka University first considered the possible exchange of electrons between protons and neutrons but found that the resulting force was too weak to provide the needed nuclear binding. Early in 1935, Yukawa proposed instead that an exchange particle whose mass of around 200 electron masses would ensure that the force would have a short range. With the resulting exchange force and with a more complex mathematical formulation than that for electromagnetism, Yukawa constructed a model for the behavior of nuclei. The new nuclear force was designed to be stronger than the well-known electromagnetic force at close distances but to decrease more rapidly as the distances increased because of the heavy particles being exchanged. At large distances the electromagnetic force would still dominate and account for the binding of atoms into molecules and molecules into solids and liquids.

Because the new particles appeared to have masses intermediate between those of the electron and the proton, they were first called mesotrons. The shorter term meson is now generally used, and in a more restricted way. Discovery of mesotrons came in 1937 when Anderson and Neddermeyer found CR tracks in their cloud chamber, which was operating in a strong magnetic field at their Pike's Peak laboratory 4,300 meters above sea level. The presence of mesons was shown by particles that had sufficient energy to pass cleanly through a 1-cm plate of platinum placed across the middle of their cloud chamber. The curvature of the track segment below the plate was markedly more tightly coiled than the segment above, indicating that the particle had lost energy in going through the plate. The curvatures of the two track segments, along with the calculated energy loss in the metal plate, showed that the particle's mass had to be around 200 electron masses. An alternative explanation, that the energy loss theory was at fault at high energies, was not entertained for long because it raised more problems than it solved.

Anderson's discovery was informally announced in Stockholm in December 1936 when he was accepting the Nobel Prize, awarded for his discovery of the positron a few years earlier. Mesons were independently discovered in 1937 by J. C. Street and E. C. Stevenson of Harvard University and by Yoshio Nishina, M. Takeuchi, and T. Ichimiya of the Institute of Physical and Chemical Research in Tokyo. (It is interesting to note that Yukawa's theory, published in Japan, took several years to receive attention in the West. Anderson and Neddermeyer made no reference to Yukawa when reporting their discovery; even Nishina and his colleagues seemed unaware of Yukawa's calculations.)

Yukawa's particles had been devised to meet the specific needs of nuclear stability. Their masses were expected to be substantially greater than those of the electrons but still much less than those of protons or neutrons, and the cloud chamber tracks were hailed as confirming this bold hypothesis when Yukawa's theory became known. But that theory also required that mesons possess other measurable properties. A chance observation gave added confirmation, but then a dramatic failure posed a serious problem.

Although the mesons had been devised to act as a nuclear glue, they were expected to be radioactive (unstable) when not confined within nuclei. This sort of behavior might seem paradoxical, but it was already known to be the case for neutrons. Many nuclei that contain neutrons are perfectly stable—witness the world around us—but free neutrons live for only around 10 minutes. The discovery of meson decay was therefore not unexpected; it came from E. J. Williams and G. E. Roberts of the University College of Wales in Aberystwyth. These two researchers took a cloud chamber photograph showing the track of a fast electron starting from the end of the track of a very slow meson (Figure 12.1). Clearly, the meson had decayed, transferring its charge to the electron. Presumably one or more neutral particles must also have been produced in the decay but had left no tracks.

The mesons that crossed the cloud chambers of Anderson and Neddermeyer, Williams and Roberts, and other researchers were generally moving faster than three quarters of the speed of light, spending around a billionth of a second in a chamber. The mesons' half-life had to be longer than this, but how much longer was not immediately clear, for the production sites of the mesons had not yet been identified.

The first experiments specifically designed to measure the lifetime of mesons were carried out by Bruno Rossi in 1939. By this time the political situation had forced him to leave Italy, and he made his measurements first at the University of Chicago (at the invitation of Compton), and then at Cornell University. Cosmic ray particles had been classified as either penetrating or soft, on the basis of their ability to pass through layers of absorbing material stacked above or inside the cloud chambers or between Geiger counters. The penetrating particles turned out to be mostly mesons, and Rossi measured their numbers at different altitudes, with and without absorber blocks above his counters. The underlying idea was that mesons, produced somehow by cosmic rays in collisions with atoms at all levels of the atmosphere, would move rapidly to lower altitudes. Along the way, their numbers would be reduced by decay and collisions. The number measured at any altitude would thus depend on a combination of those two removal mechanisms. Measurements at different altitudes, with and without extra absorber, would separate the two effects.

Rossi had to make some assumptions about the average kinetic energy of mesons. He also had to allow for an effect predicted by the special theory of relativity: because of their extremely high speeds, mesons would appear to live longer than their "true" lifetime (defined when they were measured at rest). Although this relativistic effect has now been confirmed in countless ex-

Figure 12.1. First example of a muon decay, observed in 1937 in a cloud chamber by E. J. Williams and G. E. Roberts. The muon enters the chamber at A, and travels to F, where it decays to produce a secondary electron, whose short and thinly ionizing track is barely visible. (Courtesy of L. Thomas and D. F. Falla, University College of Wales, Aberystwyth.)

periments, Rossi's demonstration was the first and is still one of the most elegant. In the end, Rossi obtained a value of about 2 microseconds for the mean life of a meson.

In a second experiment, the lifetime was measured much more directly. Electrons were detected from mesons that had slowed down and come to rest in a block of absorber surrounded by many Geiger counters. The time interval between the meson's arrival and the emergence of the decay electron was determined by a high-speed electronic circuit designed by Rossi; he obtained a mean lifetime of 2.15 microseconds—very close to the present best value.

By this time Yukawa's theory of meson properties had become known. It was expected that two versions of mesons would exist, something like the electrons and positrons with their identical masses but opposite electric charges. Traveling through the largely empty atmosphere, both positively and negatively charged mesons would decay, producing positive and negative electrons. In contrast, the behavior of mesons was expected to be quite different when they came to rest in dense concentrations of atoms. The positive particles should decay, but the negative ones should be attracted to positively charged nuclei and be rapidly absorbed, long before they could emit their customary electrons. Absorption of a meson into a nucleus might cause that nucleus to split or to emit particles such as protons, neutrons, or alpha particles. A check on these predicted meson properties was the next order of business, and it was carried out by Marcello Conversi, Ettore Pancini, and Oreste Piccioni in Rome during the period 1943 to 1945.

Conditions in Rome in those days were chaotic. Conversi had been rejected for military service because of a vision problem, and thus was able to continue intermittently with his CR research. Piccioni was at one time arrested by the Germans while trying to link up with the advancing Allied forces, but was soon released. Pancini, at one time with the partisans, rejoined the research group in 1945 after the liberation of northern Italy. Conversi, also involved with the growing Resistance movement, had moved the laboratory closer to Vatican City, where he thought it would be safer from Allied bombing.

Once things settled down, the three researchers were able to identify mesons according to their charges and look for their decay electrons. It became clear that the negative mesons were not being captured by nuclei in the way that had been expected. Nuclear captures did occur when negatively charged mesons were stopped in iron absorbers, but not when the absorber consisted of the much lighter atoms of carbon. This result, described by the Rome group in their 1945 paper in the *Physical Review,* showed that the mesons so hopefully identified earlier with Yukawa's heavy exchange particles did not possess the critical quality of always interacting strongly with nuclei. They simply could not be the agents of the strong force that binds nuclei.

The discovery of positrons and mesons came through the use of cloud chambers, with the later addition of Geiger counters to provide the coincidences for triggering the chambers. The Rome experiment had used only counters. After 1945, improved Ilford photographic emulsions provided the experimental basis for the next discoveries, made in 1947.

To study the effects of cosmic rays, Powell and Occhialini at Bristol University used some of the new emulsions, which were left at high-altitude research stations such as those on the Pic du Midi in the Pyrenees, on the Jungfraujoch in the Alps, and in the Bolivian Andes. During exposures of some weeks, numerous CR tracks were accumulated, but the early emulsions suffered from fading—older tracks did not possess as many grains as more recently formed tracks, and quantitative data were initially unreliable. But a few tracks were conspicuously different from all the others. In each case a steady increase in the closeness of grains, and an accompanying increase in scattering, identified a track as having been produced by a particle that was coming to rest. From its end another track started, and it too displayed the characteristic signs of a particle that had started at high speed, then slowed and stopped.

Grain measurements showed that each primary particle (labeled π, whence π-meson, or pion) had a mass of around 350 times the mass of an electron. The emerging secondary particle (originally labeled μ, (whence μ-meson, or muon), had a mass in the range of 200–300 electron masses. The striking feature of these early tracks was that all of the secondaries (the μ's) had the same length, close to six tenths of a millimeter. This indicated that all had started with the same initial energy, and the immediate conclusion was that only one neutral (unseen) particle could have been produced at the same time in each decay (Figure 12.2). Only in this way could momentum be conserved at the moment of decay.

Figure 12.2. Early examples of successive π–μ–e meson decays. In each of these events, only a short section of the pion track (π) is shown, leading to its coming to rest. At that point, the muon (μ) emerges with a kinetic energy of 4 MeV. After traveling for only 0.6 mm, the muon also comes to rest, and an electron (e) emerges to produce the characteristic low-grain-density track. (Photograph courtesy of Peter Fowler, University of Bristol)

Unaware of the exciting new results being obtained in Bristol, Robert Marshak of the University of Rochester and Hans Bethe of Cornell proposed in the summer of 1947 a theoretical way of reconciling Yukawa's idea of the exchange force with the Rome observations of the mesons' reluctance to interact strongly. In the "two-meson" theory of Marshak and Bethe, the Yukawa meson was the agent of the strong nuclear force and was produced in violent nuclear collisions, but it lived for only a short time, and its longer-lived secondary was thought to be the CR meson already discovered by Anderson and later studied by the Rome group. Quite independently, Shoicho Sakata and Takesi Inoue had also developed a two-meson theory in 1943, but wartime conditions prevented these calculations from becoming generally known. The English translation of their paper reached the United States only at the end of 1947.

After the Rome experiments and shortly before the discovery of $\pi - \mu$ decays, the capture of mesons by nuclei had been directly observed. Both Don Perkins at Imperial College and the Bristol group found events in their emulsions in which the characteristic track of a slowing and stopping meson ended with several short black tracks. These little "stars" were produced by the capture of mesons by nuclei in the photographic emulsions. The nature of the captured mesons was not clarified until decay events were seen later that same year. It then became clear that the negatively charged pions were indeed being captured by nuclei, in contrast to the negatively charged muons, which often decayed.

Another discovery was made in 1947 by George Rochester and Clifford Butler of Manchester University while the two researchers were examining CR tracks in their cloud chamber. Most of the tracks they observed came from particles that traveled clear across the chamber, but some interacted in a lead plate across the chamber's center. Rochester and Butler found two strange tracks that did not correspond to any of the previously known types of particle behavior. Both of these photographs, reproduced in Figure 12.3, showed tracks with abrupt bends, so the tracks looked like large Vs. One of these looked like an inverted V, with its two prongs pointing down; the other looked like a V on its side. Many CR tracks do show sharp changes in direction where the particle has bounced (scattered) off the nucleus of a gas atom.

Figure 12.3. First cloud chamber photographs showing tracks of V-particles. Left: The charged V-particle decayed at the top right, and the charged secondary was energetic enough to go clear through the lead plate across the chamber. The mass of the entering V-particle was at least 1,000 times the electron mass. Right: The neutral V-particle decays just below the lead plate, and the tracks of the two charged secondaries form the inverted V. The mass of the unseen neutral V-particle was estimated to be around 770 time the electron mass. In both photographs, other cosmic ray tracks stream across the chamber. Slow electrons can easily by identified by their strongly curved tracks. (Photographs courtesy of George Rochester.)

At the point of such a scattering there should be a track produced by the recoiling target nucleus, but none was seen by Rochester and Butler. Another distinguishing feature of these V-particles was the ionization along each track. Ionization in a cloud chamber shows up in the number of small liquid droplets along a track. Although drop counting was not as reliable as the corresponding grain counting in photographic emulsions, Rochester and Butler could still draw some firm conclusions that precluded any conventional explanation.

The measurements of these two researchers showed that the inverted V must have been produced by the decay of an uncharged particle that came out of the lead plate. Track analysis indicated that the mass of this neutral particle had to be more than about 800 electron masses, clearly heavier than Powell's recently discovered pions. In the other event, where a track moving down into the lead plate took a sudden bend, there was a marked decrease in ionization at the bend. A decrease in ionization signaled an increase in speed, but in conventional scattering a particle's speed cannot increase in this way; the particle is more likely to lose energy and emerge with a reduced speed. Treating this occurrence instead as a decay of a fast particle, Rochester and Butler proved that the primary particle's mass had to be much more than that of the pion. Other research groups began to find V-particles, and it soon became clear that the generic "V" classification actually covered several different types of particles with a variety of masses and decay schemes.

At this stage, around 1950, the gold rush was on as the number of research groups multiplied. Some set up cloud chambers on any accessible mountain; others (with emulsions) took to balloon flying. The groups at the University of Minnesota and at Bristol University were leaders in the use of balloons. In the United States balloons were flown with helium filling, but this gas was not available in any large quantity in Europe, and the flammable hydrogen had to be used—fortunately with only one explosion. International ballooning expeditions were organized for flights over Italy during 1952 to 1954, and the emulsions flown were shared among numerous European laboratories.

Here was born the first of what has become routine—enormous collaborative efforts in particle research. Recognizing the size and expense of having emulsions large enough to be able to stop even the highest-energy particle secondaries, Powell organized an international collaboration for the 1954 balloon flight and examination of the G-stack, the largest emulsions that Ilford could manufacture. The G-stack (G for gigante) results appeared in *Il Nuovo Cimento* in 1955, listing 36 authors. This journal had been started by Galileo's friends in the Accademia del Cimento in 1610, and the colophon still appears on its front pages. The G-stack paper broke new ground for the number of authors on a single paper. It is not uncommon nowadays to find the first page of a paper in *Physical Review Letters* taken up with a list of the 400+ authors and their institutions.

There was keen competition between the cloud chamber and emulsion groups as a torrent of new results came out. New particles were discovered; more accurate values were obtained for their masses; and various particles and their secondary products were finally identified. Inevitably, new terminology was invented. Mesons came to mean strongly interacting particles produced in collisions; these included pions and kaons (originally termed heavy or K-mesons), about four times heavier than pions (Figure 12.4). Kaons, some charged and others neutral, were discovered among the many V-particles. Leptons were the weakly interacting particles, themselves decay products of mesons, and included electrons and muons. Hyperons were particles that were heavier than protons or neutrons but that always produced either a proton or a neutron upon decay. For example, one of the more plentiful hyperons (first seen as an uncharged V-particle) was named the lambda (Λ) because of its appearance in cloud chambers; when it decayed after about 10^{-10} sec, it yielded only a proton and a pion. Charged hyperons, named the sigma and xi particles, were also found.

Finally, a new type of atomic nucleus was discovered by Marian Danysz and Jerzy Pniewski at the Institute for Experimental Physics of the University of Warsaw. In a photographic emulsion that the Bristol group had flown on one of their balloons, Danysz and Pniewski found a large "star" resulting from a CR collision. One of the emerging tracks was generated by a heavy nucleus, probably boron, that traveled about a tenth of a millimeter and then stopped. In turn, several other tracks started from its end, including one that might

Figure 12.4. Decay of a heavy meson (labeled τ in this 1951 photomicrograph) into three pions. The heavy track is immediately distinguishable from those of the emerging lighter pions, and its mass was calculated to be 960 electron masses. (Photograph courtesy of Peter Fowler, University of Bristol.)

have been a pion. The total energy of tracks from this secondary star was clearly so large that the boron nucleus must have possessed a remarkable amount of internal energy. Danysz and Pniewski proved that the best interpretation of this event was that a lambda-particle had been included in the nucleus of the particle that had come from the first star, replacing one of its usual neutrons. Many other examples of this sort of nucleus have since been found. These unstable hypernuclei, as they are called, have lifetimes comparable to that of the free lambda, around 10^{-10} sec.

Incorporation of a lambda particle within a nucleus introduces some exotic nuclei not otherwise found in nature. For example, hydrogen has three well-known isotopes: the dominant ^{1}H, with a single proton as the nucleus; heavy hydrogen or deuterium, ^{2}H, with a nucleus consisting of a proton and a neutron; and the radioactive tritium, ^{3}H, which has two neutrons and a proton. No ^{4}H was known until the discovery of ^{4}H$_\Lambda$, which consists of a proton, two neutrons, and a Λ (lambda) particle.

The flood of new observations revealed no initial ordering and no systematic grouping. This disarray moved J. Robert Oppenheimer to refer to the "nuclear zoo" in his talk at a 1956 international conference at the University of Rochester. With better measurements of particle masses and lifetimes, a semblance of order began to appear in the classification of the particles and their decay products, and the theorists brought their ingenuity to bear. A much deeper understanding of nuclear forces emerged, recognized through the award of Nobel Prizes in 1957 to T. D. Lee of Columbia University and C. N. Yang at the Institute for Advanced Study in Princeton, and in 1969 to Murray Gell-Mann of Cal Tech.

This exciting phase of CR research lasted only until about 1955. As the size of particle accelerators grew, so did the energies at which controlled beams of particles could be produced, and thus cosmic rays were displaced as the source of high-speed projectiles and the locus of particle studies. The first results from the new generation of machines came in 1953 from the 3 GeV protons produced in the Cosmotron at Brookhaven National Laboratory on Long Island. This was followed by the 5.6-GeV protons from the Bevatron at the University of California's Radiation Laboratory in Berkeley, where beams of kaons were produced in 1955 and antiprotons were discovered that same year.

Louis Leprince-Ringuet of the Ecole Polytechnique vividly described the coming ascendancy of accelerators at the 1953 International Cosmic Ray Conference in Bagneres de Bigorre:

> Surgit une marée, une inondation, un déluge qui s'amplifie progressivement et qui nous oblige à monter de plus en plus haut. C'est evidemment une position qui n'est pas trés comfortable, mais n'est-ce pas là une situation extrêmement vivante et d'un merveilleux intérêt?"

> A tide is rising, a flood, a deluge which is steadily growing, forcing us to climb higher and higher. Clearly this is not a very comfortable position, but isn't it a situation that is extremely lively and wonderfully interesting? (p. 290)

Today, almost 50 years later, the inundation has long prevailed, and the field of elementary particle physics has been transformed far beyond the imagination of any of the early workers. Not only have the accelerators increased in size—particles now travel in a vacuum around circles many kilometers in diameter—but so have the particle detectors. Rochester and Butler used a cloud chamber measuring 30 cm across; the UA-2 experiment at the CERN accelerator in 1981 was several stories tall and weighed more than 200 tons. States within the United States and even countries have vied for the selection of the next large machine, for the expected prestige and also for the economic benefits. It is a far cry from sending up a few pounds of emulsion with a balloon.

Is there any role remaining for cosmic rays in the study of high-energy collisions? Yes—perhaps even two. First, CR energies far exceed anything remotely possible in the laboratory. The only way to examine collisions at energies above about 10^{18} eV is still by way of cosmic rays and their extensive air showers. They come from remote and uncontrollable sources, are extremely infrequent, and are totally unpredictable in their arrival. There is also the possibility of an accidental discovery made in the course of otherwise routine experiments. A team of Bolivian and Japanese scientists studying emulsions exposed on Mount Chacaltaya have found events in which hundreds of parallel tracks appear to start from collisions outside the emulsion containers. The origin of these "Centauro" events is not yet known; they could come from the decay of a new type of particle or from a rare type of collision process.

The study of cosmic rays has made major contributions to elementary particle physics, essentially establishing it as a field of research. It is an important legacy that exemplifies the role of accident in scientific discovery and the ingenuity of scientists in improving their laboratory techniques.

13

FOOTPRINTS AND SOUVENIRS

LIKE MUCH OF SCIENCE, the pursuit of cosmic rays is remote from everyday experience. Yet cosmic rays do create effects that touch our lives. Most of these effects are produced by protons and alpha particles, the most abundant of the primary solar and galactic rays. They penetrate both the atmosphere and solid materials, where their collisions generate new isotopes—some stable, others radioactive. Sensitive laboratory techniques have enabled researchers to detect minute quantities of these isotopes and to use these CR products in two major applications: as clocks and as tracers. The present-day concentration of an isotope in a material can be used to determine the age of that material. The best-known example is dating by radiocarbon, or carbon 14, which is produced by cosmic rays in the Earth's atmosphere and enters every living animal and plant. Measurement of ^{14}C in organic remains—cloth, skeletons, wood—has revolutionized archaeological dating. Other CR radio-isotopes help to measure the ages of meteorites and the turnover rate of lunar soil by meteorite bombardment.

Cosmic rays penetrate more than the detectors on satellites and balloons. CR can and do affect computer memories. At sea level we are continually irradiated by secondary CR particles and receive an inescapable radiation dose. Such effects tend to grow in strength as altitude increases; for instance, astronauts or passengers on high-altitude planes (such as the Concorde) might receive severe radiation doses during large solar flares. In this chapter I will discuss these effects and how some of them have been put to use.

Questions of chronology have fascinated civilized society for thousands of years. How old is the Earth, and how was it formed? How and where were humans created, or when did they evolve into recognizable human form? For a long time myths and legends provided the only answers, but the acceptance of scientific methods and the discovery of radioactivity changed this. The age of the Earth is now securely based on measurements of lead isotopes and of radioactive uranium and thorium, whose relative proportions change measurably over billions of years. History and archaeology focus on the much shorter time span within which societies developed—and left numerous relics. The dating of buildings, skeletons, and artifacts was first established from a variety of clues, including historical records, chronicles of kings and dynasties, legends, and the strata of occupation at a site. When Willard Libby at the University of Chicago invented the radiocarbon dating technique in the 1940s, he initiated a revolution in historical studies.

Carbon is critical for life, for living organisms consist of a bewildering variety of carbon-based molecules. Long chains of carbon atoms, with attached loops and branches of carbon and other atoms, constitute DNA and the complex molecules needed for life and replication. The exchange of carbon dioxide between plants and animals, atmosphere and oceans, is a necessary part of the life cycle. Carbon is synthesized in stars; the carbon we find on Earth consists almost entirely of a mixture of two isotopes: 98.9 percent is ^{12}C, six protons and six neutrons in each nucleus, and the other 1.1 percent is ^{13}C, which carries an additional neutron in each nucleus. Carbon isotopes with half-lives of minutes or shorter have been created in laboratories but hold no geological or historical interest. Carbon 14, which is generated by cosmic rays in the Earth's atmosphere, is quite different.

As primary cosmic rays penetrate the atmosphere, they encounter air of increasing density, and by 50,000 feet in altitude most will have undergone nuclear collisions in which neutrons are produced. The neutrons quickly lose enough energy to be "thermalized," and their speeds are reduced to the point where they can be captured by nitrogen nuclei to produce ^{14}C.

The newly created ^{14}C is radioactive, with a half-life of 5,730 years. It decays back to nitrogen after emitting an electron. The latent radioactivity does not affect the chemical ability of the carbon 14 to combine with other atoms to form molecules. The radiocarbon combines with oxygen and produces radioactive carbon dioxide, $^{14}CO_2$, which then spreads through the biosphere. Many of these molecules attach to minute dust particles and follow air currents, becoming incorporated in rain. Some carbon dioxide is absorbed by plants that are eaten by humans and other animals; some goes into rivers and then into oceans. What makes all this possible is the long half-life of carbon 14, far longer than the typical 7 years needed to spread from cosmic rays in the stratosphere to the atmosphere, weather systems, plants and animals, and the oceans.

Carbon 14 has been produced long enough to have a steady inventory on Earth, with production just balanced by decay. There are roughly 54 tons of radiocarbon on Earth, with an average concentration of one carbon 14 atom for every trillion regular carbon atoms. This is the concentration existing in all living things. When a plant or an animal dies, its intake of carbon dioxide with its carbon 14 ceases. Thereafter, the $^{14}C/(^{12}C+^{13}C)$ ratio changes slowly as carbon 14 decays. Determining this ratio enables us to estimate the age of the wood, skeletons, paper, or cloth.

Measuring such low concentrations of carbon 14 is not a simple process. In the first method, devised by Libby, electrons from the carbon decay in a small sample are counted. Detecting the electrons requires a combination of particle counters to distinguish them from the larger numbers of particles coming from local radioactivity and sea-level cosmic rays. Because the objects to be dated are usually unique and testing destroys any sample taken, there is understandably great pressure to keep the sample sizes small; one tenth of a gram (about 1/300 ounce) is usually the largest available. This effectively sets a limit to the age that can be determined in this way, for in very old materials most of the radiocarbon will have decayed, and the counting rate from a tiny sample will be lower than background levels.

The radioactive signal of the equilibrium concentration of radiocarbon shows up in the emission of 13.5 electrons per minute for each gram of carbon in a fresh sample. In wood that is 2,000 years old, for example, about 79 percent of the radiocarbon will have survived and the decay rate will be about 10 counts per minute. By 4,000 years, the corresponding rate will be nearly 8 counts per minute.

A far more sensitive technique for carbon 14 detection was developed in the 1970s; it identifies atoms by their masses rather than by their radioactivity. A minute sample is vaporized and the fully ionized atoms are sent through a cyclotron or other type of accelerator. The high-speed nuclei emerge to travel through well-calibrated magnetic fields in which the curvature of their paths depends on their speed, mass, and charge. Counters, similar to those used in CR work, measure their energies and speeds, and the arrival of each nucleus is individually noted. In this accelerator mass spectrometry (AMS), only a few milligrams of sample are needed, and the carbon 14 can be clearly separated from the more abundant carbon 12 and 13.

Converting measured carbon 14 concentrations to ages is not a straightforward process. "Dates" derived from radiocarbon measurements must be corrected for variations in the CR's production rate. There will also be variations in the diffusion rate of the radiocarbon. Still, with the most sensitive techniques, ages of up to 70,000 years can now be determined under favorable circumstances.

From measurements of the radiocarbon in the annual rings of bristlecone pines and sequoia trees, radiocarbon calibration has been constructed and the basic technique validated. Growth patterns exhibit a long-term drift between carbon and "true" dates, with additional short-term variations. Wood samples whose rings are no more than 3,000 years old tend to have ages overestimated by the radiocarbon technique, while the effect is reversed for older samples. The size of this effect is generally in the 10 to 15 percent range; thus wood whose rings are 7,000 years old will be seen, by radiocarbon, to be about 600 years younger. One cause of this difference is the slow drift in the strength of the Earth's magnetic field, which controls the entry of the low-energy CR that produce the radiocarbon. Another factor is the variation of CR intensity during the 11-year solar cycle.

In a paper delivered at the 1995 International Cosmic Ray Conference, G. E. Kocharov, of the Physico-Technical Institute of the Russian Academy of Sciences in St. Petersburg, Russia, gave a very clear demonstration of this cyclic variation in carbon 14, as measured in pine trees in the South Urals, for the years 1600–1730 (Figure 13.1). This covers the period of the Maunder Minimum, when sunspots were virtually absent and more of the lower-energy primary cosmic rays could enter the Earth's atmosphere. It is these cosmic rays whose collisions lead to the production of the carbon 14 (see Chapter 4).

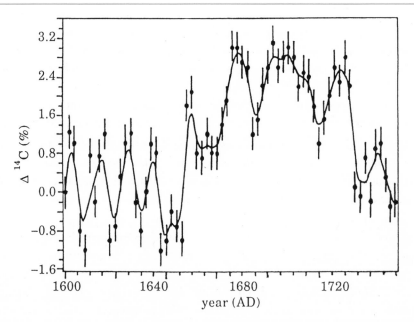

Figure 13.1. Carbon 14 variations in the biennial rings of pine trees in the South Urals, 1600–1730. Data points are displayed, together with a curve obtained by interpolation and smoothing. Production of ^{14}C increased after 1645 when solar activity was low; see Figure 4.1. (Courtesy of G. E. Kocharov, Physico-Technical Institute, Russian Academy of Sciences, St. Petersburg, Russia.)

Even with the calibration from tree rings, caution is still needed in interpreting the measurements. For example, apparent ages of 2,000 years have been found for some living organisms whose carbon dioxide had come from limestone on the floor of a freshwater lake, which, in turn, was deficient in radiocarbon.

Serious problems affecting radiocarbon dating have been created by humans. The burning of large quantities of fossil fuels, begun in the nineteenth century, has injected enough low-^{14}C carbon into the atmosphere to decrease the radiocarbon ratio by about 3 percent. Even worse, before the atmospheric testing of nuclear weapons ceased in the early 1960s, so much radiocarbon had been produced and dispersed that the atmospheric inventory was doubled. The overall result of these alterations and various other uncertainties is that future calibrations cannot be based on today's wood, and dates within the past 200 years are hard to establish with reliability. But the tree-ring calibration does still show that, to within better than 50 percent, CR intensity has been fairly constant over the past 8,000 years, and in general the radiocarbon method is still of great use.

The tracking of radiocarbon is effectively confined to the Earth because of the atmosphere and the life cycle. Many other radioisotopes are produced by both galactic and solar CR, on the Earth and on the moon, in sufficient quantities to be of interest. The longest-lived of these are listed in Table 13.1. Beryllium 10 (^{10}Be) is particularly interesting because of its use as a chronometer for CR travel in the galaxy. Most ^{10}Be in the solar system is thought to have been

Table 13.1 Long-lived isotopes produced by cosmic rays.

Isotope	Half-life (years)
^{10}Be	1.5×10^6
^{26}Al	7.4×10^5
^{36}Cl	3.0×10^5
^{40}K	1.3×10^9
^{53}Mn	3.7×10^6
^{59}Ni	7.5×10^4
^{129}I	1.6×10^7

produced by cosmic rays, some in space and rather less by collisions in our atmosphere. Its half-life is long enough for it to be transported from the atmosphere to the surface, where it is found in sea water, polar ice, igneous rocks, ocean sediments, and nodules on the ocean floor. Another isotope produced by cosmic rays, ^{26}Al, with a half-life close to three quarters of a million years, has chemical behavior similar to that of beryllium. The Be/Al ratio is useful as an index for determining how fast atmospherically produced isotopes are carried through geological processes. In addition, their measured abundances show that the cosmic rays that generated them must have had roughly the same intensity for the past 5 to 10 million years.

On the Earth, wind, rain, and vegetation combine to eliminate or mask the evidence of old meteorite impacts, but on the moon those surface features last indefinitely—or until the arrival of another meteorite or an astronaut. The samples retrieved by the *Apollo* astronauts during the 1969–1972 missions established an absolute chronology for some lunar surface features. Before that time only a relative chronology could be constructed; for example, in a place where debris from one crater lay across another feature (perhaps another crater), we knew the sequence of formation but not how long ago the events had occurred. Now, through the application of radiodating methods to lunar samples, we know that the moon's age is about 4.6 billion years and that most of the major cratering ceased by 3 billion years ago. Since then there have been occasional violent impacts with the arrival of large bodies and a steady drizzle of much smaller objects, ranging in size from small rocks through micrometeorites of millimeters or smaller down to individual nuclei from the solar wind and from solar and galactic CR.

Small meteorites can make microscopic pits in lunar rocks; larger ones carve small craters in the lunar surface and throw out debris. Because of the low surface gravity and absence of air resistance on the moon, this debris will travel farther than it would on Earth. The lunar surface is thus continually being churned up, with fresh debris overlying that from earlier impacts. The net result is a "gardening," with deeper layers turning over less rapidly because they require the infrequent arrival of larger meteorites.

With no shielding atmosphere to impede them, primary CR penetrate the lunar surface and produce radioisotopes, as they do on Earth. CR penetration naturally decreases with depth, so material closer to the surface receives a larger dose of cosmic rays than does deeper material. To this must be added the gardening effects, in which fresh (less irradiated) material is brought to the top by cratering, where it is exposed to a larger radiation dose than it otherwise would have received.

In addition to producing radioisotopes, cosmic rays are responsible for causing radiation damage on a different scale, within certain types of crystals. Some energy lost by cosmic rays is stored within the crystals and can be retained for a long time. Subsequent heating of the crystals releases the stored energy, which appears as light. This thermoluminescence (TL) reveals the radiation dose that the crystal received and thus the number of CR particles that went through the crystal. TL monitors are widely used as industrial dosimeters, routinely worn by people who work near X-rays and radioactive materials.

During some *Apollo* missions, astronauts hammered hollow tubes more than 2 meters into the lunar surface to collect core samples that preserved the relative positions of the different layers. From the TL in these cores, the CR exposure at different depths and thus the gardening history could be calculated. In this way, it was found that the top half millimeter of the lunar surface is turned over 100 times in a million years. At a depth of 1 cm, 50 percent of the material is untouched in that period of time; it takes about 10 million years to turn over the top centimeter. The turnover time scale lengthens to a billion years between 10 cm and 100 cm.

Measurement of TL, even in small crystals, demonstrates a bulk effect in that the energy is accumulated throughout each crystal from the passage of many cosmic rays over many years. The effect of an individual CR particle is not measured. Tracks of individual cosmic rays have been seen, however, through the process of chemically etching suitable materials. Figure 13.2 shows such tracks, which were revealed by etching a crystal from a lunar rock with sodium hydroxide. During the *Apollo* missions, some detectors were deployed on the moon's surface and promptly brought back. One of the more striking demonstrations of CR tracks was seen in a filter from the camera on the Surveyor 3 lunar lander. This unmanned vehicle had landed on the Moon in 1967, and the filter was retrieved by the *Apollo 12* mission over 2 years later.

When the solar system formed by the contraction of a great cloud of gas and dust, most of the condensed mass ended up as the sun, with less than 1 percent as much in the planets and their satellites. It was long thought that most of the original cloud was somehow expelled from the solar system. A tiny fraction of the original mass remains in the form of comets, meteorites, and dust. For many years, comets could be studied only by remote observation of their often spectacular but generally unpredictable appearances. A first step

Figure 13.2. Tracks of cosmic rays, revealed after etching of a crystal from a lunar rock. Each of these tracks is less than 1/100 mm in length. (Photograph courtesy of Robert Walker, Washington University, St. Louis.)

toward laboratory-based experiments was taken when the 1985 and 1986 space probes intercepted the periodic comets Halley and Giacobini-Zinner, whose returns can be accurately forecast. Plans are being made to intercept future comets, collect samples, and return them to Earth.

The study of interplanetary dust started with Earth-based observations of the gegenschein and the zodiacal light, sunlight reflected from dusty regions in the solar system. From spectroscopic analysis we know that submillimeter dust grains are responsible for these phenomena and that the solar system contains a great many other dust grains, for example from comet tails. Dust particles have been collected by high-flying airplanes and identified, by their composition, as certainly not being terrestrial in origin. These interplanetary dust particles (IDPs), typically a few hundredths of a millimeter in diameter and weighing around a billionth of a gram, have been examined with a battery of laboratory techniques at the very limits of sensitivity. Beyond showing that the composition does not match with terrestrial material, the measurements have revealed a record of CR irradiation (Figure 13.3).

In many of the dust particles, CR tracks have been observed in the electron microscope after chemical etching. Track densities as high as 10^{11} tracks per square centimeter have been seen, for example, in a grain of the mineral olivine ($MgFeSiO_4$). With present-day CR intensities, it would take only about 10,000 years to accumulate this number of tracks. Observation of these tracks also tells us that during interplanetary travel and even atmospheric entry, the grains could not have been heated above roughly 500° C, for at higher temperatures the tracks would have been obliterated.

Another indicator of the extraterrestrial origin of the grains is the detection of relatively large quantities of those isotopes of the rare gases helium, neon, argon, krypton, and xenon that can be produced only by CR interactions. Many techniques being applied to the IDPs were developed over the years as the status of more easily obtained meteorites shifted from museum curiosities to prized scientific specimens.

Ages for many meteorites have been determined by the application of radiometric techniques using isotopes of lead and uranium and an isotope of rubidium that has a half-life of 49 billion years. This method was previously used for dating terrestrial and lunar samples. "Age," in this context, means the time since the meteorite solidified enough to retain the gaseous radioactive daughter products that would otherwise leak away. In this way, ages up to 4.6 billion years have been deduced.

Meteorite "CR ages" have also been found from analyzing CR tracks in

Figure 13.3. An interplanetary dust particle, around three thousandths of a centimeter in diameter. Measurement of the isotopic composition of hydrogen in such particles has demonstrated their extraterrestrial nature and has also provided information on their bombardment by cosmic rays during their interplanetary travels. Tracks have been found in some of these particles. (Photograph courtesy of Robert Walker, Washington University, St. Louis.)

mineral grains, under the assumption that the tracks accumulated throughout the time when the meteorite surface was exposed to cosmic rays at the present-day intensity. CR exposure ages of 5 to 50 million years have been found for some stony meteorites, and rather greater ages (200 to 1,000 million years) for some iron meteorites. These ages correlate with the mean lifetimes for these types of meteorites in orbits that cross those of the Earth and Mars; the CR ages suggest that these objects might have originated in the breakup of a group of larger bodies millions of years ago. The picture is one of steady meteorite generation: fresh surfaces are formed that then accumulate CR tracks over times quite short relative to the age of the solar system.

Other isotopes, such as ^{39}Ar and ^{31}Cl, are produced by cosmic rays inside the meteorite after it has landed on the Earth's surface. Terrestrial "ages" (since impact) from 270 years to 1.5 million years have been measured through the use of these isotopes.

Roentgen's dramatic picture of the bones in his hand immediately caught the attention of the public and the medical profession. Since then, for more than a century, X-rays have been used for diagnostic purposes and have become an indispensable tool of modern medicine. Before the health hazards of X-rays were recognized, X-ray machines were widely used in shoe stores, and children enjoyed seeing the skeletal images of their wiggling toes as shoes were being fitted. X-rays have also been used therapeutically, especially in the treatment of some types of cancer. If these were the only ways in which the word radiation turned up in everyday use, we might well have few worries, but Hiroshima and Nagasaki changed that. To what extent do cosmic rays produce unwanted and adverse effects?

Many physical processes used to detect cosmic rays also occur when CR and other high-speed particles travel through living tissue. Some of the particles' kinetic energy is transferred to the tissue, where electrons can be ejected from atoms. In this way chemical bonds can be broken and the disrupted molecules can serve as the focus for cancer to develop. Radiation is used to kill cancer cells, but healthy cells can also be killed.

Different systems of units describe radiation doses. In the older system, the rem (roentgen equivalent in man) was defined to correspond to the absorption of around 100 ergs of energy per gram of tissue. (The qualification "around" refers to the differing biological effectiveness of different forms of radiation such as protons, neutrons, X-rays, and electrons, which are included in a more specific definition.) By international agreement, a newer unit, the sievert (sv) has been adopted, with 1 sv equal to 100 rem. But because so much of the popular literature still uses the older units, I will do so as well.

The average radiation dose that every person receives each year is about one tenth of a rem, or 100 millirem (100 mr), derived from three roughly equal sources. The contribution of the cosmic rays varies somewhat with geographical latitude and longitude; this radiation dose is distributed evenly throughout the body. A second contribution, delivered internally and concentrated in some organs, comes from the radioactive potassium 40 that is part of our natural chemical makeup. The third source is the average daily medical dose (X-rays and nuclear medicine) received by the population, at least in developed countries. Therapeutic medical doses will generally be much larger, but those are not received by the general population and are applied only when there is a clear medical reason and when benefits outweigh risks.

Inhalation of radon, a gaseous daughter product from radium, may expose many people to an additional 100 mr, mainly in the lungs, but this dose varies widely across the country, depending on factors such as local geology, type of building material, and degree of ventilation in housing. The potential hazard posed by excessive radon inhalation is widely recognized, and commercial kits can be bought for easy monitoring.

A lethal dose of radiation, such as was received in 1945 by people in Hiroshima and Nagasaki, exceeds 400 rem, but the CR dose rate for an individual, even accumulated over a lifetime, is far lower. On the other hand, the total CR dose received annually by the entire U.S. population is 270 million × 30 mr or about 8 million rem. The incidence of cancer is in the range of one case for every 10,000 rem; so perhaps 800 cases of cancer are unavoidably induced each year in the United States by cosmic rays. This is far lower than the total annual cancer rate and thus should not be a cause for great concern—there is not anything we can do about it and we have no way of identifying the people whose deaths have come from their CR doses.

Potentially more serious is the dose that could be received by airplane passengers and crew and astronauts who are traveling at high enough altitudes to be exposed to large numbers of solar CR during major flares. The giant flare of February 1956 would have produced a peak dose rate of 100 mr per hour in people flying at 35,000 feet, far above the usual commercial airplane altitudes of those days but quite normal today. During a single transatlantic flight, a passenger could thus receive a dose at least equal to the normal yearly accumulation. Crew members in a supersonic plane, traveling at around 60,000 feet for 500 hours each year, will accumulate a dose of 500 mr from galactic CR. This amount is well above the normal population dose and is equal to the recommended annual maximum for industrial radiation workers. To avoid large doses during a solar flare, an airplane would have to reduce altitude rapidly and gain protection from the greater thickness of the overlying atmosphere. This defensive maneuver is not available to astronauts, who could encounter additional doses in passing through the Van Allen radiation belts. In the center of the belts, a dose rate of 20 rem per hour could be received. This can be avoided by keeping spacecraft orbits below 400 km or above 6,000 km.

The unseen hazard presented to astronauts by cosmic rays has been vividly illustrated in two ways. During the Earth-moon stages of their journeys, several *Apollo* astronauts reported seeing bright flashes of light, even when their eyes were closed. The flashes were described as "pinpoints that disintegrated rapidly" and "frequently streaks, with occasional double flashes." The nature and frequency of these flashes are consistent with their production by cosmic rays. Some of these CR (probably heavily charged particles) produced Čerenkov radiation in the eyeball, while others might have hit the retina directly.

There is no way that the astronauts' eyes can be examined for residual evidence of CR traversal, but support for this explanation of the flashes comes from particle tracks in the plastic helmets worn by the astronauts. As shown in Figure 13.4, etching of the plastic has revealed tracks from heavy CR, identified by their track dimensions.

If cosmic rays can cause light flashes in astronauts' eyes, they can also effect changes in the brain, but we have no evidence yet for harmful effects from single particles. Computer memories might be similarly affected, and the chance

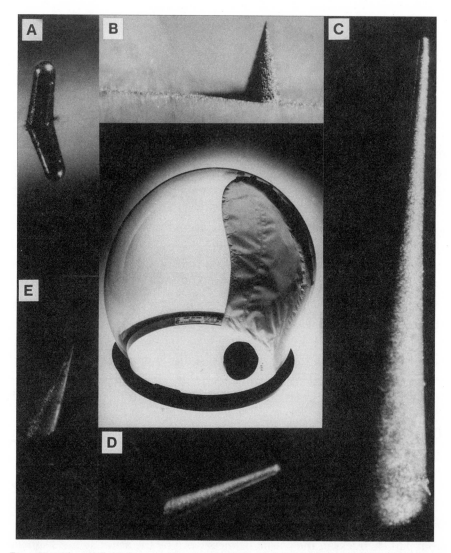

Figure 13.4. The *Apollo 8* helmet used by astronaut James (Jim) Lovell. Cosmic ray tracks in this and other plastic helmets were revealed by chemical etching and replicas are shown in the surrounding photographs. Tracks A, D, and E were found in the helmets of *Apollo 12* astronauts Charles (Pete) Conrad and Richard (Dick) Gordon; tracks B and C were in a control helmet exposed to the primary cosmic rays. Track C was 0.7 mm in length, and the others all in the range 0.35–0.60 mm. (Photograph courtesy of R. L. Fleischer et al, *Science* 170, 11 December 1970, pp. 1189–1191; © 1970 by A.A.A.S.)

for this increases during long-term satellite flights. Some types of computer problems are termed soft fails, in which a single digit changes without command, and testing reveals no way of duplicating the event. In contrast, hard fails are the result of faulty but replaceable components. The passage of a CR particle through a computer chip can trigger soft fails, and the question is whether the frequency of these fails poses a hazard. The conclusion drawn from several studies is that only heavily ionizing particles generally represent a potential threat, but the problem will become more serious as more compact computer devices come into use. It is possible, however, to incorporate error-correcting codes to guard against random errors so that computer instructions are not incorrectly interpreted.

14

EPILOGUE

THERE IS A RECURRENT THEME in scientific discovery, well described by the aphorism "one person's noise is another's signal." What is an intrusive and unwanted effect in some experiments may turn out to be of great importance in its own right. In just this way, the discovery of cosmic rays was accidental, a by-product of an attempt to understand an effect (electric charge leakage) that had emerged as a nuisance during experiments with radioactivity. Establishing the extraterrestrial nature of the new radiation and deciding among competing interpretations of the data took close to 25 years, largely because of the unrecognized role of the Earth's magnetic field and the complexity of the secondary effects in the atmosphere. Indeed, those effects were more than just complex; they resulted from the interplay of processes not even remotely imagined at the time.

The further accidental discovery of the positron and mesotron opened up the field of elementary particle physics. For 20 years, until the construction of the large accelerators in the 1950s, new elementary particles could be found only through cosmic rays—always, of course, by accident. In those early days, each new arrival was given its own name and the relationships (if any) between them were not discerned. We had the K and theta and tau heavy mesons and, among the hyperons (those particles heavier than protons), there were lambda, sigma, and xi particles. The particle zoo was more heavily populated than anyone had imagined.

Gradually, with better measurements and the ingenuity of theorists, order was brought to the scene. This provided the stimulus for the construction of larger and larger accelerators to produce particle beams of higher and higher energies with which to probe deeper into the subnuclear world.

As complementary information came flooding in from optical and radio astronomy, the 1948 discovery of the heavy primary particles broadened the astrophysical base that remains the central theme of cosmic ray studies. Jansky's discovery of cosmic radio waves led to recognition of the link between the particle and e-m radiation. With the advent of X-ray and gamma ray astronomy, which required very high altitude observations, we entered the era of high-energy astrophysics, the current and more comprehensive designation. Along the way progress was made through the development of new types of particle and radiation detectors, along with the theory of particles and their interactions, and the launching of the space age.

The scope of cosmic ray physics has continued to widen. A survey of the annual number of scientific publications on cosmic rays shows the following figures: In 1924 there were 9 papers published on cosmic rays; in 1933 there were 184; in 1950 there were close to 400. In 1960 and 1970 the numbers increased to approximately 500 in each of those years, and in 1999 there were about just over 4,000 papers with "cosmic rays" in their titles. Although cosmic ray results are reported at a wide variety of scientific meetings, the premier gathering of the clan is at the biennial International Cosmic Ray Conference (ICRC). There is even a new journal, *Astroparticle Physics*.

There are still important measurements waiting to be made. For example, more data are needed so that the fine details of the charge spectrum can be examined and the abundances of individual isotopes can be determined. As shown by the recent discovery of the abundances of the Co and Ni isotopes, the injection and acceleration stages seem to be more complex than had been thought. The measurement of the energy spectrum, especially at the highest energies and of the high-Z nuclei, must continue so that researchers can construct more refined models for the origin and propagation of cosmic rays. Further searches are needed to pin down the abundances of the antinuclei or to set upper limits. Many observations require space-borne hardware. On the ground, research on extensive air showers will continue, and ballooning will also proceed, for its own interest and also for testing prototype experiments that can later go up on satellites.

High-energy particle physics grew out of the cosmic ray discoveries, and it is interesting to note that cosmic ray observations are now providing a defense for a new accelerator project. The Relativistic Heavy Ion Collider (RHIC) is designed to produce TeV (10^{12} eV) beams of gold and lead nuclei and have them collide with equally heavy target nuclei. Some fears have been expressed that, in collisions of such violence, uncontrollable consequences could include the production of black holes or space-time warps, in which our Earth might be swallowed up. It is therefore comforting that theorists have pointed out that this experiment has already been running—for millennia—with cosmic rays. Equally energetic heavy CR particles have been slamming into the moon's unprotected surface—and the moon is still there. Moreover, no trace of anything of this bizarre kind was found in the samples brought back to earth by the *Apollo* astronauts.

The history of science has shown how research fields flourish and then dwindle or become absorbed as new fields emerge. In cosmic ray studies, we have seen this happen with the elementary particle research, while the astrophysical side has exploded. What will cosmic ray research cover at the turn of the next century? We cannot tell—perhaps a direct connection will be found with human genetics, for all of us are continually receiving a cosmic radiation dose, even if small. No matter how it turns out, those of us who have had the good fortune to participate in this research have had so much enjoyment that we should not begrudge others the excitement of seeing their fields prosper and move ahead. It is this sense of novelty and continuity, consolidation and discovery, that makes scientific research such a wonderful enterprise.

Finally, the most fitting commentary on cosmic rays was made by J. J. Thomson and G. P. Thomson in 1928, long before the full range of this exciting subject was known. In *Conduction of Electricity through Gases,* (p. 12), they wrote: "It would be one of the romances of science if these obscure and prosaic minute leakages of electricity from well-insulated bodies should be the means by which the most fundamental problems in the evolution of the cosmos had to be investigated."

WORKS CITED

BIBLIOGRAPHICAL NOTE

ACKNOWLEDGMENTS

INDEX

WORKS CITED

Baltrusaitis, R. M., et al. 1985. *Nuclear Instruments and Methods,* vol. A240.

Drake, S., trans. 1957. *Discoveries and Opinions of Galileo.* Garden City, N.Y.: Doubleday.

Leprince-Ringuet, L. 1953. "Congrès International sur le Rayonnement Cosmique." Mimeo, University of Toulouse.

Millikan, R. A. 1926. *Physical Review* 27:360.

Shklovsky, I. S. 1960. *Cosmic Radio Waves.* Cambridge, Mass.: Harvard University Press.

Störmer, C. 1955. *The Polar Aurora.* Oxford: Clarendon Press.

Taylor, A. M. 1967. *Imagination and the Growth of Science.* New York: Schocken Books.

Thomson, J. J., and G. P. Thomson. 1928. *Conduction of Electricity through Gases,* 3rd ed. Cambridge: Cambridge University Press.

Thrower, N. J., ed. 1981. *The Three Voyages of Edmond Halley in the Paramore.* London: The Hakluyt Society.

Young, C. A. 1898. *The Sun.* New York: D. Appleton.

Most of the literature on cosmic rays is in research journals or monographs and is accordingly technical. The most recent popular treatment is the book by Roger Clay and Bruce Dawson both of the University of Adelaide, *Cosmic Bullets* (Reading, Mass: Helix Books, 1997). This is a very accessible introduction to cosmic rays, with strong emphasis on their own research area of the highest-energy particles. Articles on cosmic rays appear in *Scientific American* from time to time and can be located by consulting the cumulative index. Useful compilations also appear in *Mercury,* a popular magazine published by the Astronomical Society of the Pacific.

There are some older popular treatments, now rather dated, having been bypassed by the flood of information in recent years or by the shifting balance of interests within cosmic ray physics. Two are definitely worthy of mention, although both are out of print. The first is *Cosmic Rays* by Bruno Rossi (New York: McGraw-Hill, 1964). This book is a classic by one of the cosmic ray pioneers. Rossi made major contributions to the field from 1924 on, and the High Energy Astrophysics Division of the American Astronomical Society now awards an annual Rossi Prize. The second book is Martin A. Pomerantz's *Cosmic Rays* (New York: Van Nostrand Reinhold, 1971), an eloquent survey. Pomerantz's research was mostly concerned with the solar-terrestrial aspects of cosmic rays, and his book is particularly strong in its treatment of these topics.

The discovery and early studies of cosmic rays are included in a popular history of physics by Emelio Segre, *From Atoms to Quarks* (San Francisco: W. H. Freeman,

1980). Segre won a Nobel Prize for the first observations of antiprotons, produced at the University of California's Bevatron. A similar treatment, but with many more illustrations, is *The Particle Explosion* by Frank Close, Michael Marten, and Christine Sutton (New York: Oxford University Press, 1987). This book is one of the best popular surveys of the history of particle physics.

A few other books deserve mention. *The Early History Of Cosmic Rays,* edited by Y. Sekido and H. Elliot (Dordrecht: Reidel, 1985), contains personal recollections by many of the early researchers. *The Birth of Particle Physics,* edited by Laurie M. Brown and Lillian Hoddeson (Cambridge: Cambridge University Press, 1983), contains papers presented at a conference and includes recollections from several of the pioneers of CR particle physics, including Carl Anderson, Bruno Rossi, and Marcello Conversi. Another set of recollections appears in the proceedings of a conference held in 1987 to celebrate the fortieth anniversary of the discoveries of pions and V-particles; *Forty Years of Particle Physics* is edited by B. Foster and P. H. Fowler (Bristol and Philadelphia: Adam Hilger, 1988). An extended treatment of particle history, with considerable theoretical detail, is given by Abraham Pais in *Inward Bound* (Oxford and New York: Oxford University Press, 1986*).* In *Origins of Magnetospheric Physics* (Washington, D.C.: Smithsonian Institution Press, 1983), James A. van Allen provides an account of his discoveries made with the use of rockets and satellites.

For more technical discussions, a good starting place is *Cosmic Rays and Particle Physics,* By T. K. Gaisser (Cambridge; Cambridge University Press, 1990). Other books at this level include *Introduction to Ultrahigh Energy Cosmic Ray Physics* by Pierre Sokolsky (Reading, Mass.: Addison-Wesley, 1989) and *High Energy Astrophysics* (vol 2, 2nd ed.) by Malcolm Longair (Cambridge: Cambridge University Press, 1992.) A much older classic, worth noting, is *The Origin of Cosmic Rays* by V. L. Ginzburg and S. I. Syrovatskii (New York: Pergamon Press, 1964).

ACKNOWLEDGMENTS

It is my great pleasure to acknowledge the help that I have received in many ways in the preparation of this book. Within the cosmic ray community and over many years, more colleagues than I can list here have provided stimulation and friendship that I treasure. Many of them responded generously to my requests for photographs or diagrams or advice. I am especially grateful to Rosemary Fowler for her permission to use photographs that Peter Fowler provided many years ago. The manuscript has benefited greatly from the careful scrutiny and excellent editing of Nancy Clemente at Harvard University Press. Mary Ellen Geer was of similar help with the earlier version. To all these individuals, and especially to my family for their unfailing interest and support, I am deeply grateful.